)12

Bernard Hodges is Senior Lecturer in Computer Aided Engineering at the Bournemouth and Poole College of Further Education.

Dr Paul Hallam is Deputy Director of the Technology Support Unit at Sandwell College of Further and Higher Education.

The cover picture shows a series of Cincinnati T^3 700 electric robots spot welding on an automobile production line (by courtesy of Cincinnati Milacron).

Industrial Robotics

Bernard Hodges and Paul Hallam

Heinemann Newnes

Heinemann Newnes
An imprint of Heinemann Professional Publishing Ltd
Halley Court, Jordan Hill, Oxford OX2 8EJ

OXFORD LONDON MELBOURNE AUCKLAND SINGAPORE
IBADAN NAIROBI GABORONE KINGSTON

First published 1990

British Library Cataloguing in Publication Data
Hodges, Bernard
  Industrial robotics.
  1. Industrial robots
  I. Title   II. Hallam, Paul
  629.892

ISBN 0 434 90782 0

Typeset by Vision Typesetting, Manchester
Printed and bound in Great Britain by
Butler & Tanner Ltd, Frome and London

Contents

Preface

This book has been written for students, production engineers, applications engineers and industrial managers who need a grasp of the basic principles of construction and operation of industrial robots, the uses to which they can be put in manufacturing industry and the measures necessary for their safe, economic and effective use.

The scope and level of the book are considered suitable for students of robotics and automation systems on courses ranging from BTEC NIII and City & Guilds 230 (Robot Technology and Control) to HND/HNC and first degree level.

Since the first concept of the use of robots as production tools in the 1950s, robot technology has advanced rapidly as a result of innovations in the various areas of engineering that contribute to robotics: electronics and computing, mechanical and production engineering, electrical engineering and control systems. Today, robots can perform tasks that, only a few years ago, would have been considered far beyond their capabilities and that make use of such sophisticated process elements as lasers, plasma and vision systems.

It is not possible to cover here in detail every aspect of industrial robotics, but we hope to dispel any remnants of mystique associated with the 'humanoid' robot of fiction, and provide the reader with an insight into the practical and useful applications of that developing range of production tools that is encompassed by the term 'industrial robots'. The first part of the book explains the working of the robot as a machine – its mechanical elements, drive systems and sensors. We then go on to discuss how robots are programmed, their performance specifications and the crucial subject of safety measures in installation and operation.

There follows an important chapter on economics, cost-effectiveness and the organizational aspects of installing a robotics system. We conclude with a review of the main types of industrial application of robots – with examples of successful installations – and finally a brief discussion of areas of research into advanced systems which are going to determine developments in the last decade of the century.

B.C.H.
P.H.

April 1990

Acknowledgements

The authors are grateful to the following individuals and companies who have provided information and photographs. Without their help, this book could not have been written.

ABB Robotics Ltd
Automatix International UK Ltd
Autotech Robotics Ltd
Bournemouth and Poole College of Further Education
British Robot Association
BYG Systems Ltd
Centa Transmissions Ltd
Cincinnati Milacron
A Corkhill (ARASA)
Crompton Parkinson Instruments Ltd
R. Egginton (Department of Trade and Industry)
Evershed Robotics Ltd
GMF Robotics Corporation
GMFANUC Robotics (UK) Ltd
Kim Goh (Loughborough University)
Barry Goodwin (Seres (Marketing) Ltd)
Harmonic Drive Ltd
Mike Hudson (Lightguards Ltd)
IBM Ltd
KUKA Welding Systems and Robots Ltd
Precision Systems
Reis Robot Ltd
600 Group Colchester Lathe Company
Mike Skidmore (Altec Automation)
Staubli Unimation Inc

1 Introduction

1.1 What is a robot?

Early ideas of robots and robotics

The term 'robot' was originally used to refer to an automated humanoid machine, although in the world of science and technology the word has had a much wider application. Some authors of science fiction have called their 'living' humanoids robots, while in industry the term is applied to automated systems ranging from the simple to the very complex.

During the early 1920s robot-like machines were made and used in Greece, Italy and some other European countries. The word 'robot' comes from the play 'R.U.R.' ('Rossum's Universal Robot'), written in 1920 by the Czechoslovakian playwright Karel Čapek. In the Czech language the word 'robota' means statute labour, that is, compulsory labour. In Čapek's play, Rossum and his son develop anthropomorphic machines as servants for humanity.

It appears that the science fiction writer Isaac Asimov was the first to use the word 'robotics' to describe robot technology. In his stories Asimov actually anticipated real developments and problems, some of which are possibilities that have still to materialize.

Asimov proposed three 'laws' which for many years were recognized as principles in the use of robotics:

1 A robot must not harm a human being or, through inaction, allow a human to come to harm.
2 A robot must always obey human beings unless this is in conflict with the first law.
3 A robot must protect itself from harm unless this is in conflict with the first or second law.

The development of the industrial robot

With the passing of time and the development of various new kinds of machines, robots came to be defined and classified in different ways. Two early developments are still in use today.

Manipulators

Manipulators are mechanically controlled devices for use in hazardous areas. They usually take the form of clamps or pincers mounted on the end of an arm and controlled by mechanical linkages from a remote point

(Figure 1.1). The manipulator is situated behind a wall or screen that protects the operator from the hazardous material being handled. The screen can take the form of a window so that the operator can see the operation being carried out.

In the early days of atomic and nuclear experimentation, manipulators were used to handle radioactive materials; the protective windows had to be made of heavily leaded glass several centimetres thick to absorb the radiation.

When manipulators were controlled by human operators, considerable skill was required. The introduction of servo power amplified electric motors greatly increased the power of the manipulator while making the job much easier for the human controller. The fitting of servo amplifiers and electric motors also increased the lifting and handling capacity of the manipulator.

Teleoperators

The term 'teleoperator' was coined by Edwin Johnson in about 1966 to describe servo controlled manipulators that are not directly connected to their control systems. Remote control devices are used to operate teleoperators by means of joysticks, wheels or control keyboards. The control systems use radio links, optical links or sometimes satellite communications systems.

Such devices often require a form of vision, especially when the device is a long way from the control – this may be provided by remote television cameras. Other types of sense may be required, such as touch, which initially was based on a feedback system working through the servo system. Development in this type of sensing has now progressed to the transmission of radio links.

Teleoperators were developed mainly for hazardous operations such as undersea work, work with radioactive materials and support systems for space operations. They are still in use today in the USA for handling radioactive materials.

Classifying robotic systems

Robots can be classified according to their application, their configuration or their load capacity. Perhaps the easiest way to envisage the main robot types is to group them according to their applications:

Materials-handling robots
Assembly robots
Welding robots
Paint-spraying robots

These applications can be expressed in more technical terms, such as degrees of freedom, payloads, repeatability.

However, the various types of robot can be distinguished more clearly by

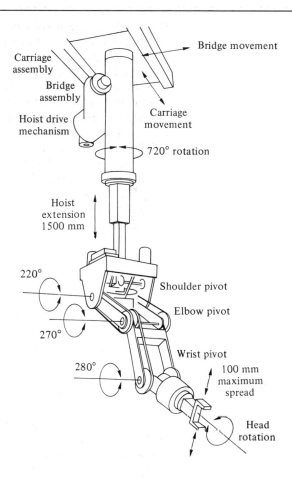

Figure 1.1 Manipulator

classifying them according to their operating principle:

Cartesian coordinate
Cylindrical coordinate
Polar coordinate
Articulated-arm coordinate
Gantry
SCARA

These coordinate systems are discussed in Chapter 2.

The pattern of development since 1960

The modern robot was developed by Joseph Engleberger and George Deroe, whose work led to the formation of Unimation Inc., early manufacturers of industrial robots. The first industrial robot application, in 1961, was a press-loading operation. Unimation sold industrial robots to many companies – notably General Motors – but robots did not become a major force in industry generally until they had been used extensively in the Japanese automobile industry.

Figure 1.2 The ASEA robot deburring the oilways on a crankshaft (*Courtesy ABB Robotics Ltd*)

Figure 1.3 FANUC M series robot working in conjunction with a machining centre and conveyor (*Courtesy GMFANUC Robotics (UK) Ltd*)

Looking back over the three decades leading to the 1990s, the 1960s may be categorized as the decade in which numerical control came to maturity; the 1970s established the computer and, in particular, CNC (computer numerical control) and the 1980s saw astounding advances in the development of robot control, robot sensing and the quality of robot manufacture.

The result has been increased acceptance of the industrial robot worldwide, bringing a notable improvement not only in productivity but also in the vital sphere of quality control. Standards for product quality are increasingly emphasized and are embodied in the UK in BS 5750. Some companies link quality control standards with Just-In-Time procedures (JIT) and manufacturing resource planning (MRP) for production control and a smoother operating system.

Figures 1.2 to 1.5 show some of the applications of industrial robots in the late 1960s.

Figure 1.4 (*above*) Polishing 90° stainless steel elbows, the first industrial application of an ASEA robot
(*below*) An ASEA robot polishing stainless steel kitchen sinks (*Courtesy ABB Robotics Ltd*)

Figure 1.5 The ASEA IRB 6
laying down adhesive (*Courtesy
ABB Robotics Ltd*)

1.2 Considerations in introducing robotized systems

Since robots began to be used in industrial applications there has been much debate as to the real value of robotized systems in the production and manufacturing sectors: the viability of high investment to introduce unproven benefits of automation has been questioned. Moreover, it has not been unusual for robots to be purchased as the 'solution' to a problem without proper identification of that problem. Companies have bought robots because it was fashionable to do so, without the necessary feasibility studies and preparation; this resulted in robots lying idle in the corners of factories, collecting dust – either because there was not enough work for the robot to do or because the wrong robot was purchased for the particular application.

Feasibility studies

The applications that industrial robot users are implementing today are much more realistic than they were ten years ago. Companies tend to complete their own feasibility studies in-house, highlighting the components or processes that may be worked on by industrial robots. With this established, the advantages of purchasing such capital equipment can be worked out thoroughly before embarking on serious discussions with robot suppliers or turnkey manufacturers.

The feasibility study must have the full backing and support of senior managers, middle management and production engineers. Total commitment to the project must stem from the shop floor upwards to ensure successful implementation.

Main points of a feasibility study

- Analysis of the product or process to ascertain whether it is suitable for automation using robots
- Study of the cycle times to determine whether they are long enough to warrant the use of a robot
- Selection of the right robot to perform the task with the required accuracy and repeatability
- Selection of a supplier who will have full turnkey accountability
- Analysis of how the parts are to be presented to the robot
- Design of work-holding devices to be as flexible as possible, enabling variations of product within the product family to be catered for
- Investigation of the necessary safety standards

Motivating factors

The motivating factors that affect the introduction of a robotics system can be categorized as technical, economic and social.

Technical factors

When comparing robot and human performance it is generally considered that humans cannot match the speed, quality, reliability, endurance and predictability of robotics systems. However, robots cannot compete with hard automation if the cycle times are short and flexibility is not an important factor – for instance, robots would not be considered for the manufacture of paper clips or washers. Robots therefore provide a link between the rigidity of dedicated automation and the flexibility of the human operator, in that they offer:

- High flexibility of product type and variation
- Lower preparation time than hard automation
- Better product quality
- Fewer rejects and less waste than labour-intensive production

Economic factors

The overriding consideration in the application of robotized systems is the associated financial commitment and prognosis. Major factors in considering the possible implementation of robotics systems include:

- The need to increase production rates to remain competitive
- Pressure from the marketplace to improve quality
- Increasing costs
- Shortage of skilled labour

The redesign of manufacturing lines may present the opportunity of manufacturing new product lines.

In general, robotics can increase profitability by:

1 providing maximum utilization of capital-intensive production facilities for up to 24 hours per day, seven days per week;
2 reducing production losses due to absenteeism and skilled labour shortage;
3 reducing the amount of inventory which is being processed with resulting savings in work in progress (WIP);
4 reducing the manufacturing lead time of the product or processes;
5 reducing scrap and increasing product quality, with resulting reduction in the number of customer complaints.

Social factors

The introduction of any type of automation into a company is undoubtedly going to have some impact on personnel, both on the shop floor and at management level. The belief that robotics systems cause a reduction of labour with resulting unemployment has become less widespread with developing understanding of the role that robots play, and the growing appreciation that people can be relieved from mindless, repetitive tasks to utilize their skills in more interesting work or to be retrained for different work. Many low-level tasks can be carried out by robots, as can undesirable work in dangerous or hazardous environments and work requiring heavy physical effort. This has been shown clearly in the automobile industry where robots are used for spot-welding car bodies, applying spray paint and underseal and for the application of sealants and adhesives that give off harmful vapours. In the nuclear industry telemanipulators and robots have been used successfully for carrying out maintenance work on reactors and for the handling of dangerous waste products.

At the other extreme of robot capability, robots are being integrated with vision systems for carrying out inspection, fitted with tactile sensors to handle the most delicate of components and are mounted on tracks or other forms of slide to give them greater mobility.

In general, workers on the shop floor do not object to the introduction of robotized techniques since they appreciate that their tasks are going to be made easier, with resulting increased productivity and with the removal of dangerous and physically demanding conditions. Also organized labour has learned to accept that the adoption of new technology is a critical factor in the fight to preserve jobs, by expanding market share and maintaining the competitive position of products. In considering the cost of remaining competitive, it may be noted that labour costs more than doubled during the 1980s; in the same period technological capability – as measured by processing speed, memory capacity and feature improvements – increased by more than four times.

When considering the introduction of a robotized system it is imperative that the workforce is informed from the outset that certain work may be taken over by a robot. Not only will the worker then be involved in the planning, but he/she will be able to give valuable guidance on how the task can best be performed to maximum productivity. Greater job satisfaction can be gained from working with robotic systems and in fact the image of the company can be enhanced by implementing new technology. However, it is important to take into account from the outset that robots cannot contribute the flair and imagination of the human operator.

1.3 Statistics

It is difficult to give significant figures relating to the number of robots installed throughout the world, their application and related capital expenditure, since these statistics are not easily obtainable and are very quickly out of date. Some indication of this data can be obtained from information published by robot associations in various countries.

The British Robot Association (BRA) was the first body to publish figures relating to the number of robot installations and other relevant data some years ago and these statistics, which are updated annually, tend to serve as a benchmark for other associations. The purpose of this section is not to load the reader with statistics and numerical information but to outline trends and new developments indicated by the figures released by the BRA.

Figures 1.6 and 1.7 show the numbers of robots installed in different industrial sectors to December 1988 and the applications of robots installed during 1988. The total was 5034 robot installations, an increase in robot population of 17% during that year. As expected, the automobile industry retained the largest robot population, followed by the rubber and plastics sector. Of the 731 robots installed during 1988 the largest application was machine loading, accounting for 31.7% of the total. This is not surprising since machine loading and unloading are repetitive, can be hazardous and can easily be performed by a relatively simple and inexpensive robot. This is borne out by the fact that of the 731 robots installed, 36% were valued at between £10 000 and £20 000. Injection moulding and arc welding together represented 31% of the total.

This highlights some of the changes in trends in the UK and in opinion as to how and where robots should be used. It is interesting to note a decline in robots used for assembly operations, i.e. 58 installed during 1988 compared with 86 in the previous year. All robot manufacturers recognize that this area holds the highest potential for sales, but it may be that the technology has not yet developed sufficiently to enable dexterous assembly tasks to be executed faster and more reliably by robots.

The robot cost analysis published by the BRA shown in Figure 1.8 shows the cost breakdown of robots installed in the UK during 1988 in relation to the country of manufacture. It is noteworthy that, of the robots of UK origin, most were at the lower end of the cost scale, i.e. 153 robots costing

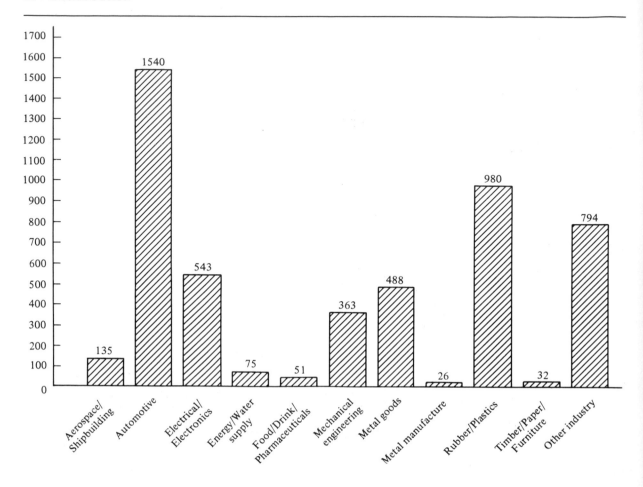

Figure 1.6 Total number of robots installed in the UK to December 1988 (*Courtesy British Robot Association*)

between £10K and £20K, with very little activity in the more expensive range. In contrast, Japan and the EEC had greater sales of robots costing over £20K; these are obviously more sophisticated robots and the figures indicate that it would not necessarily be profitable for these manufacturers to export lower-cost machines. Although Japan supplied 60 robots in the range £20K to £30K, these must be compared with robots manufactured in the West costing over £35K: Japanese robots sell at around two-thirds the price of similar European and American robots.

To complete the basic statistics for robot installations during 1988, Figure 1.9 shows a technical and cost analysis of installed robots within the price range £10K–£35K, classified as either servo driven or non-servo driven. It can be seen that robots used for injection moulding and machine loading applications, relatively simple tasks, represented the greater number in the non-servo, low-cost range. Servo-driven arc-welding robots dominate those costing over £20K. The largest number of robots costing below £10K were in education/research and assembly applications. This may well be an indication of training robots being purchased in the education sector and low-cost SCARA or simple pick-and-place robots for automating assembly tasks.

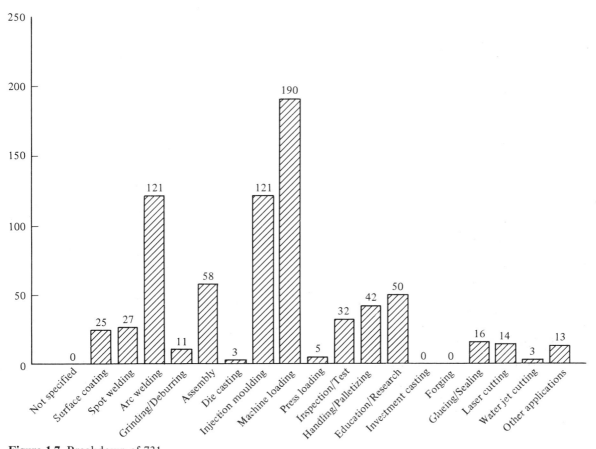

Figure 1.7 Breakdown of 731 robots installed in the UK during 1988 (*Courtesy British Robot Association*)

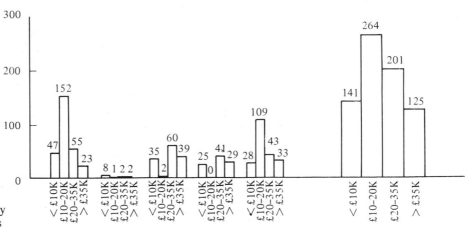

Figure 1.8 Cost analysis by country of origin of robots installed in the UK during 1988 (*Courtesy British Robot Association*)

Application	Robot cost in £1000				Technical	
	< 10	10–20	20–35	> 35	non–servo	servo
Surface coating	4	3	0	18	4	21
Spot welding	2	0	1	25	2	25
Arc welding	12	0	74	35	12	109
Grinding/Deburring	5	0	3	3	5	6
Assembly	36	1	18	3	15	44
Glueing/Sealing	10	4	1	1	5	11
Laser cutting/Welding	0	12	2	0	14	0
Water jet cutting	0	2	1	0	0	3
Die casting	2	0	1	0	2	1
Injection moulding	0	115	6	0	75	46
Machine loading	14	113	42	21	118	72
Press loading	2	1	2	0	3	2
Inspection/Test	1	8	19	4	1	31
Handling/Palletizing	8	2	25	6	15	26
Other applications	0	1	4	8	0	13
Education/Research	45	2	2	1	1	49
Totals	141	264	201	125	272	459

Figure 1.9 Technical and cost analysis of robots installed in the UK in 1988 (*Courtesy British Robot Association*)

2 Mechanical elements of robots

In this chapter the various mechanical elements of robots are outlined, such as different types of robot configuration and the working envelopes that they produce. The methods used to drive the axes of manipulators in both linear and rotational directions are discussed. The chapter concludes with a section on robot end-effectors, their design and operating principles, with typical industrial examples.

2.1 Robot coordinate systems

As outlined in Chapter 1, over the years robot manufacturers have developed many types of robots of differing configurations and mechanical designs, to give a variety of spatial arrangements and working volumes. These have evolved into six common types of system: articulated arm, polar, cylindrical, cartesian, gantry and SCARA (Selective Compliance Assembly Robot Arm).

Articulated-arm configuration (RRR)

All the movements of joints or arm members are rotary, as shown in Figure 2.1: θ is the horizontal rotation about the base, ω is the vertical angular motion about the base and U is the vertical angular movement of the arm relative to the first arm member. β, α, and γ are the angular movements of the wrist, which can be in either two or three axes depending on the customer's specification. Other examples of articulated-arm robots are the Puma robot, which was designed to duplicate the size and configuration of a human arm, and the FANUC S100 robot. These robots are illustrated in Figures 2.2 and 2.3.

Polar configuration (RRT)

Two of the joints move in rotation and the third moves in translation (Figure 2.4a). An example of this type of configuration is the Unimate robot shown in Figure 2.4b.

Cylindrical configuration (TTR, RTR, RRT)

Figure 2.5 shows a cylindrical robot; the principal movements are 300° rotation about the base and translation or linear travel in the vertical and horizontal planes. Figure 2.6 shows a typical cylindrical or pick-and-place robot manufactured by Cincinnati Milacron.

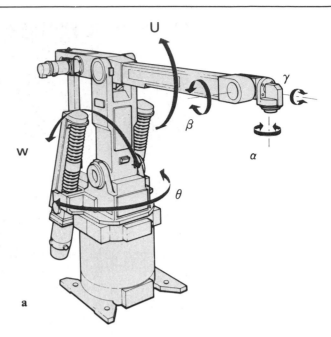

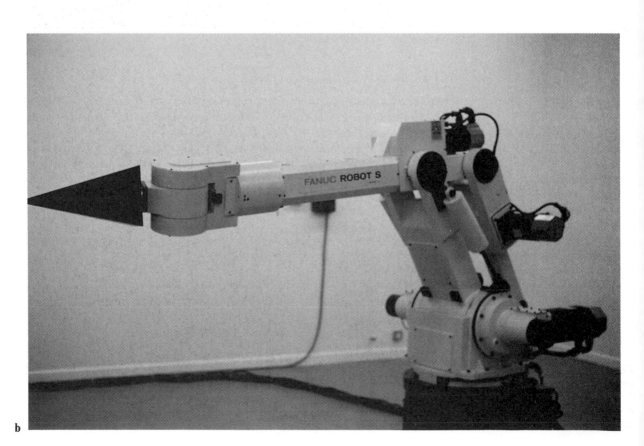

Figure 2.1 Articulated-arm robot (*Courtesy GMFANUC Robotics (UK) Ltd*)

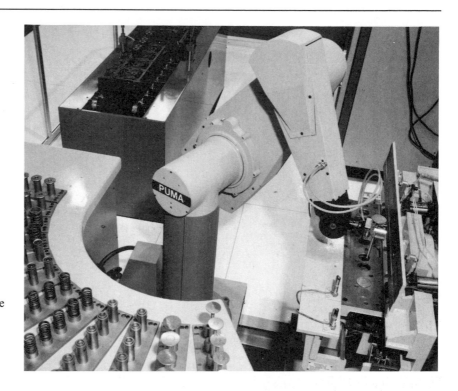

Figure 2.2 A Unimate Puma 560 Mk II robot engaged in the assembly of a 6-cylinder head complete with twelve valves, springs and cotter insertion (*Courtesy Staubli Unimation Ltd*)

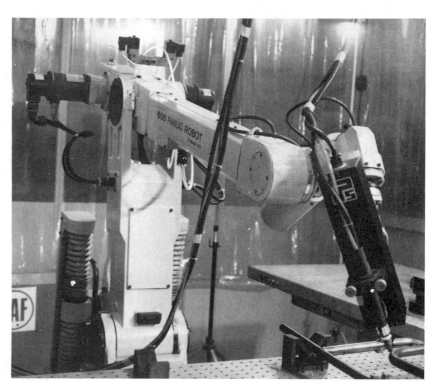

Figure 2.3 FANUC S100 articulated-arm robot pulse TIG welding (*Courtesy Bournemouth and Poole College of Further Education*)

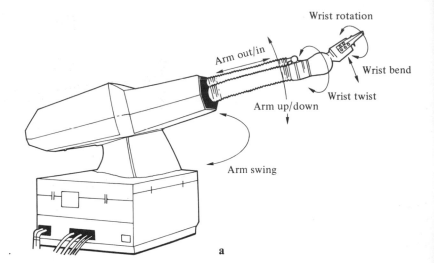

Figure 2.4 a The 6-axis system of the polar configuration **b** A polar-type robot, with vacuum-gripper loading, packaging grates (*Courtesy Staubli Unimation Ltd*)

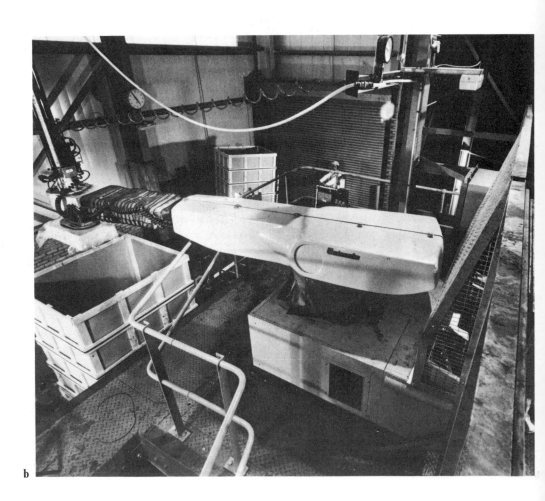

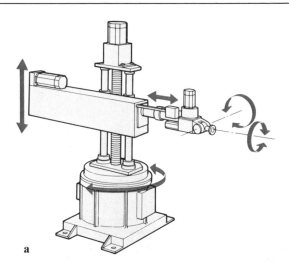

Figure 2.5 Cylindrical robot configuration (*Courtesy GMFANUC Robotics (UK) Ltd*)

a

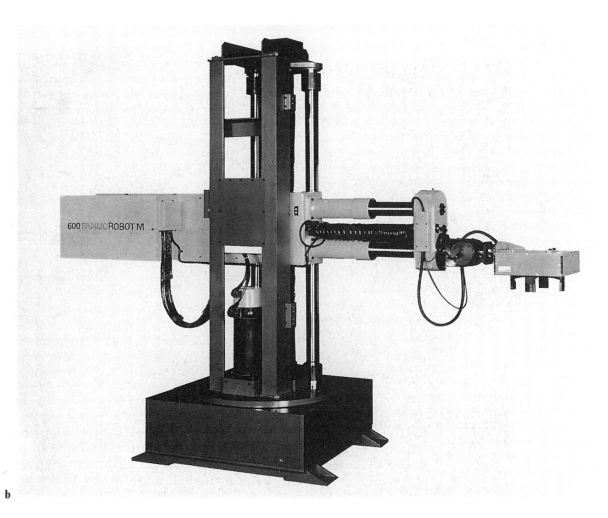

b

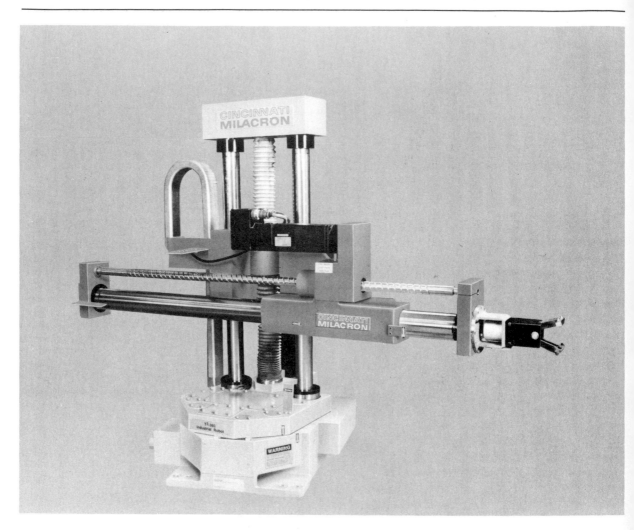

Figure 2.6 The Cincinnati
Milacron T³ 363 (electric)
(*Courtesy Cincinnati Milacron*)

Cartesian configuration (TTT)

These robots move in three directions in translation at right angles to each other. Figure 2.7 shows the axis movement. The Cartesian robot system is sometimes referred to as the linear configuration.

Gantry configuration (TTT)

This configuration has the same freedom of movement as the Cartesian but includes an additional support to give rigidity. This is illustrated by the KUKA IR robot shown in Figure 2.8. An industrial application of this type of robot is shown in Figure 2.9 where a Reis electric robot is depalletizing and loading timber boards into woodworking machinery.

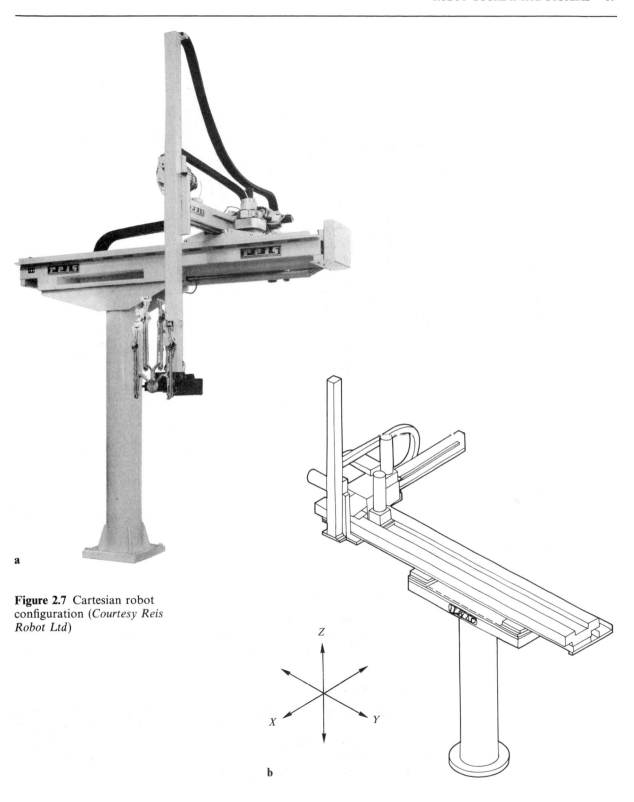

Figure 2.7 Cartesian robot configuration (*Courtesy Reis Robot Ltd*)

Figure 2.8 A gantry-type robot
(*Courtesy KUKA Welding Systems and Robots Ltd*)

Figure 2.9 A Reis electric gantry-type robot depalletizing and loading timber boards into woodworking machinery
(*Courtesy Reis Robot Ltd*)

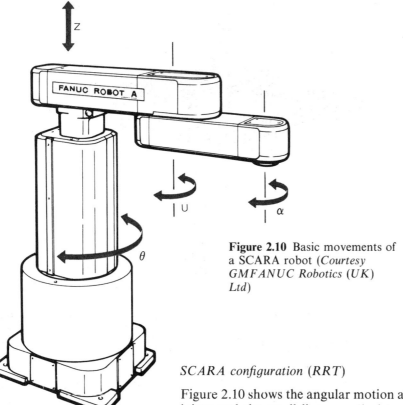

Figure 2.10 Basic movements of
a SCARA robot (*Courtesy
GMFANUC Robotics (UK)
Ltd*)

SCARA configuration (RRT)

Figure 2.10 shows the angular motion about the base, the first and second joints and the small linear vertical travel. These robots were developed primarily for assembly tasks; they are very fast and have a repeatability of better than 0.025 mm. Figure 2.11 shows a typical example of two 600 FANUC SCARA-type robots assembling printed circuit boards.

Degrees of freedom

A degree of freedom can be defined as the direction in which a robot moves when a joint is actuated; usually each joint represents one degree of freedom. Some robot manufacturers designate the opening and closing of the gripper as one degree of freedom, but it is not generally recognized as such. The majority of robots in use today use five or six degrees of freedom, but the choice is very much dependent upon the robot task or application: for example, a welding robot requires five or six degrees of freedom, but for a pick-and-place application only three axes may need to be specified.

The degrees of freedom of a robot depend upon its configuration. If we consider an articulated-arm type, the six axes are:

1 The vertical movement up and down in the Z direction.
2 The extension and retraction of the arm in the X direction – this effectively changes the radius of the manipulator.
3 The rotational traverse in the θ direction about the base of the robot or in the Y direction when operating in the Cartesian mode.

Figure 2.11 Two FANUC SCARA robots assembling printed circuit boards (*Courtesy GMFANUC Robotics (UK) Ltd*)

There are also the three degrees of freedom on the wrist of the robot:

4 The wrist pitch, the β axis, which is the up or down movement of the wrist.
5 The yaw, or γ axis, which is the movement of the end-effector to the left or right.
6 The wrist roll, or α axis, which is the rotation of the wrist clockwise or counterclockwise. The amount of rotation is dependent on the make of robot: some have the capability of turning through 20 turns, the limiting factors being the wrist drive mechanism and the number of wraps that the cabling can withstand.

These degrees of freedom are illustrated in Figure 2.12 and these movements are sufficient to place the end-of-arm tooling in any position or orientation within the robot's working envelope.

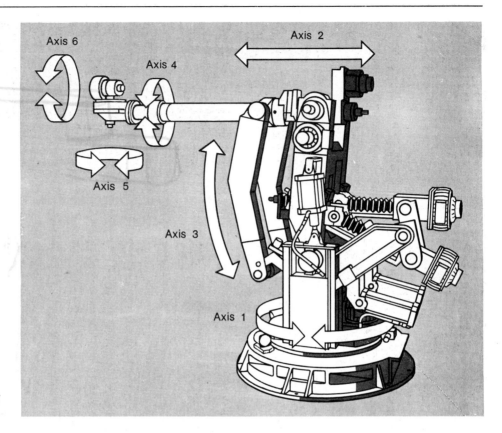

Figure 2.12 Degrees of freedom

Working envelopes

The working envelope of a robot is defined by the points that can be reached by the maximum and minimum movements of each axis or of a combination of axes. The envelope or volume generated that is quoted in manufacturers' literature is normally that taken from the mounting flange at the end of the arm. It is quoted from this position because it is the purchaser who selects the appropriate end-of-arm tooling, which may be bought from another company or made in-house. Selecting the robot with the appropriate working envelope for the task to be carried out is of great importance, as all interaction with peripheral devices and processes must take place within this generated volume. Besides having the capability to reach the equipment within the operating volume, the robot must avoid collisions with workpieces and with other robots working in its immediate vicinity.

For each robot configuration, a different size and shape of operating space is generated. In fact, two robots of the same configuration can have widely differing working envelopes resulting from the size and type of arm design. Although the arm configurations of the articulated-arm robots shown in Figure 2.13 are similar, the working envelopes generated are quite different. This type of robot has a high ratio of robot size to working volume which allows its application to a wide range of tasks.

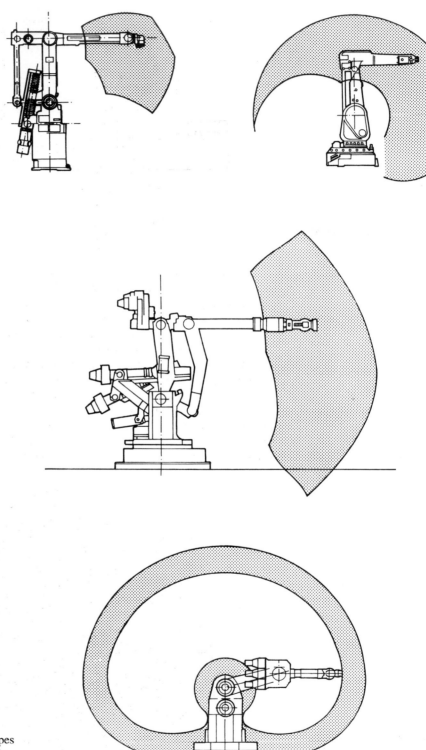

Figure 2.13 Working envelopes of four articulated-arm robots

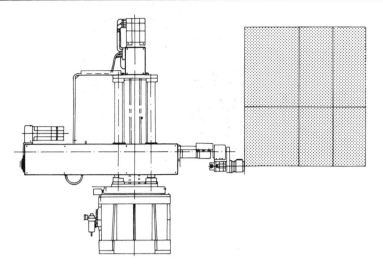

Figure 2.14 Operating envelope of a Cartesian robot

The rectangular envelope of a Cartesian robot, shown in Figure 2.14, makes this type of robot ideally suited to materials handling and machine loading. Figure 2.15 shows the operating envelope of a 600 FANUC SCARA-type robot, where the linear movement in the Z direction is small in relation to the other axes since a large up-and-down movement is not needed for assembly operations.

The working envelope of a robot can be changed by altering the mounting arrangements. This is shown in Figure 2.16, where a KUKA robot is depicted in four different mounting positions: overhead, floor-mounted and at angles of 45° and 90°. Similarly, the robot can be mounted on rails or tracks, so creating a seventh axis or degree of freedom. This increases the size of the operating volume and adds to the flexibility of the robot since it may greatly reduce the need for dedicated component transfer

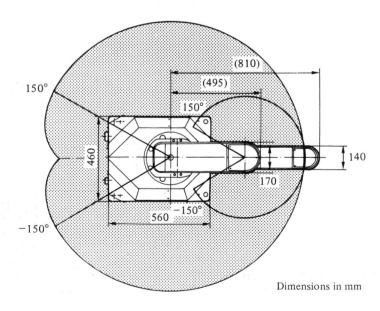

Figure 2.15 Operating envelope of a SCARA-type robot

Dimensions in mm

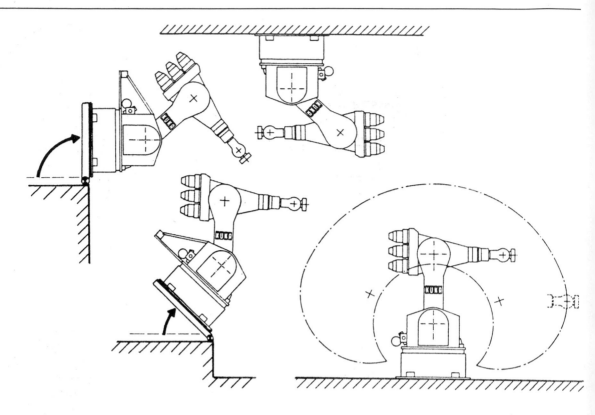

Figure 2.16 Different mounting arrangements change the working envelope (*Courtesy KUKA Welding Systems and Robots Ltd*)

systems and special feeders. This is illustrated in Figure 2.17 which shows the increased working envelopes of KUKA IR 361/8/15 and IR 361/8/15.2 robots when a seventh degree of freedom is added.

The motion track on which the robot moves is supplied by the robot manufacturer; an example of such a track is shown in Figure 2.18. The track shown is servo-controlled and is programmed direct from the robot controller. It can carry up to 300 kg, excluding the robot, for 11 metres and may be either floor-mounted or hung overhead. An industrial application of a robot operating along a track is shown in Figure 2.19 where a Reis R625 40 kg six-axis robot is running on the seventh axis to implement the cleaning of process machinery in the tobacco industry.

In 1984 ASEA introduced the IRB 1000 6-axis pendulum robot for assembly tasks. This was quite revolutionary since the robot arm was suspended in the direction of gravity and the masses of the arm were concentrated around the gyratory point. This resulted in a robot that has high precision and repeatability, i.e. 0.1 mm, and is capable of moving very fast, in the order of 12 m/s. The concentration of mass at the pivot point results in a low moment of inertia that enables very high accelerations to be achieved. The vertical robot arm is mounted in a stiff frame that allows the arm to pitch and roll like a pendulum with three axes of motion at the wrist. The working volume of this type of robot is quite different from those mentioned previously, since the volume generated forms a combination of a sphere and rectangle (Figure 2.20).

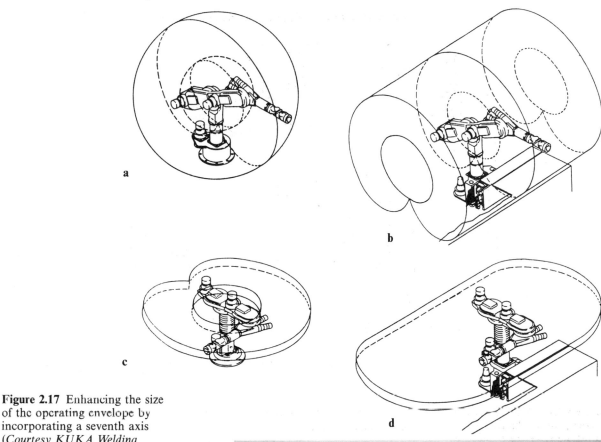

Figure 2.17 Enhancing the size of the operating envelope by incorporating a seventh axis (*Courtesy KUKA Welding Systems and Robots Ltd*)
a Vertical swivel, stationary: RRR–RRR. With carriage as seventh axis: TRRR–RRR
b Vertical swivel, mobile: TRR–RRR
c Horizontal swivel, stationary: TRR–RRR
d Horizontal swivel, mobile: TTRR–RRR

Figure 2.18 (*facing page*) Robot mounted on a servo-controlled linear track (*Courtesy ABB Robotics Ltd*)

Figure 2.19 Robot operating on a track to clean process machinery (*Courtesy Reis Robot Ltd*)

Advantages and limitations of different robot configurations

Cartesian

As Cartesian robots move in three linear directions it is easy to comprehend where the arm is in space. This not only makes the human task of programming easy but also reduces the amount of computer power required. Due to their rigid structure these robots can manipulate high loads so they are commonly used for pick-and-place operations, machine

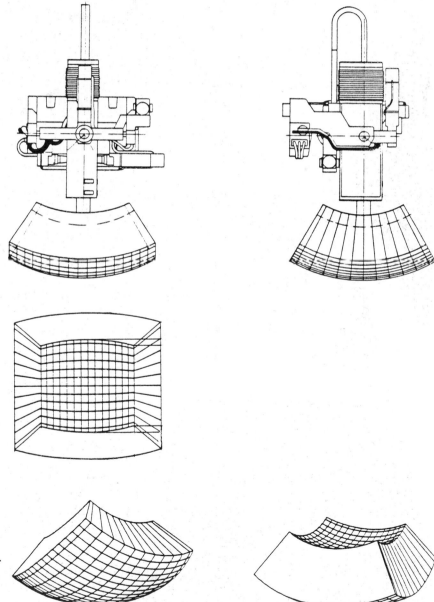

Figure 2.20 Working volume of the ASEA IRB 1000 pendulum robot (*Courtesy ABB Robotics Ltd*)

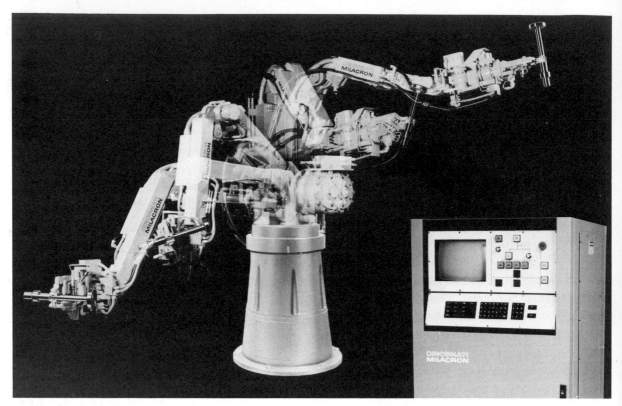

Figure 2.21 Cincinnati Milacron T³ 566 heavy duty robot (*Courtesy Cincinnati Milacron*)

tool loading, stacking parts in bins, in fact any application that uses a lot of moves in the *X*, *Y*, *Z* planes. They are also used for assembly, e.g. electronics parts insertion – a major growth area.

These robots do, however, occupy a large space, giving a low ratio of robot size to operating volume. Also all three axes are vulnerable to environmental dust and dirt so they may require some form of protective covering. The same applies to gantry-type robots.

Cylindrical

As with Cartesian robots, it is easy to visualize where the end-effector is going to be as there is only one rotary axis of movement. Cylindrical robots have a rigid structure, giving them the capability to lift heavy loads through a large working envelope, but they are restricted to areas close to the vertical base or the floor. This type of robot is relatively easy to program for loading and unloading of palletized stock, where only the minimum number of moves is required to be programmed.

Polar

These are fairly versatile robots, possessing one linear and two rotating axes that can generate a large working envelope. This configuration has been used by Unimation Inc. since the early 1960s for spot welding and loading die-casting machines. The robots can be powered by electricity or can be hydraulically driven to allow large loads to be lifted. As there are two rotary movements it is more difficult to visualize and control the end-effector position than it is for Cartesian or cylindrical robots. The semi-spherical operating volume leaves a considerable space near to the base that cannot be reached. This design is used where a small number of vertical actions is adequate: the loading and unloading of a punch press is a typical application.

Articulated arm

This is perhaps the most widely used arm configuration because of its flexibility in reaching any part of the working envelope. All joints are rotary, which adds to the complexity of the computer system and makes it almost impossible for the operator to visualize the end-effector path or position. Robots such as the Cincinnati Milacron (Figure 2.21) and the Thorn EMI Workmaster hydraulic robot are capable of lifting loads well in excess of 150 kg. In the case of electric drives, the AC or DC motor is completely sealed so enabling the robot to operate not only in harsh environments but also in clean-room conditions. This configuration flexibility allows such complex applications as spray painting and weld sealing to be implemented successfully.

SCARA

These robots are very fast and accurate, having speeds of the order of 3 m/s with repeatability of ± 0.02 mm. The operating envelope is formed by two or three links driven by electric motors; these generate a series of curves which enable the end-effector to move through 360° in the plan view and reach the whole of the working volume. Although originally designed specifically for assembly work, these robots are now being used for welding, drilling and soldering operations because of their repeatability and compactness. They are intended for light to medium loads and the working volume tends to be restricted as there is limited vertical movement.

2.2 Robot drive mechanisms

As already discussed in Section 2.1, the axis moves of a robot can be either linear or rotary and in some cases a combination of the two is used. These movements are achieved by using various mechanical devices to convert the rotation of, say, an electric motor into a precisely controlled linear arm movement. Some of the most common types of mechanism are now outlined.

Lead screws

A precision screw rotates, driving along its thread a pre-stressed nut which is attached to the driven part. This type of drive is sometimes used on CNC machines but is seldom used on robots because of the lack of rigidity and the high friction forces involved.

Ball screws

Ball bearings are fed out of carrier into a precision-ground recirculating ball screw for three or four revolutions and then fed back into the carrier. This produces a low-friction drive since the load is being transmitted by point contact rather than by sliding contact as in the case of a lead screw. This type of drive is virtually free of backlash because the minimal contact between the balls and the screw greatly reduces wear between the two mating parts.

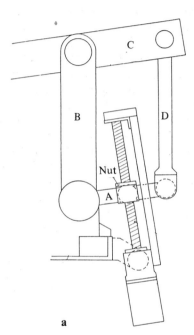

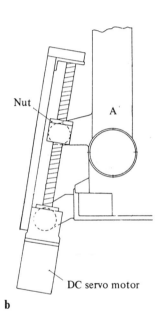

Figure 2.22 Drive mechanisms for an articulated-arm robot (*Courtesy GMFANUC Robotics (UK) Ltd*)
a Arm movement
b Shoulder movement

Most articulated robots use ball screws to convert the rotational motion of a DC motor to drive the shoulder and the arm axis, as shown in Figure 2.22. In Figure 22a the DC servo motor drives the ball screw which moves the nut up or down. The nut is attached to the link A which forms the input side of a parallelogram consisting of links A, B, C and D. As the ball-screw nut moves up or down, the ball-screw motion is transmitted from the lower link A through link D to the arm C. Figure 22b shows the drive to the shoulder. The DC servo motor drives the ball-screw nut which is attached to the link A; the motor is pivoted on the base, so when the nut moves up or down the link rotates about its pivot point.

Chain and linkage drives

Chain drives are used on some robots to transfer the rotary motion of an electric motor to a sprocket-driven joint situated at the end of the robot arm. The advantage of this type of drive is that locknut tensioners can easily be incorporated into the chain to take up slack in the chain or resulting from wear on the sprocket or wrist drive. As an alternative to chain drives, ASEA robots have used linkage drives to transmit the motion generated by the electric motors mounted in the base or the shoulder. These consist of discs, attached to the motors, which drive push-pull rods; these in turn are fixed to rotating discs, situated at the joint, which produce the required motion.

Gear drives

Gear drives are used to transmit and change the angular velocity, direction and torque of the output shaft relative to the input. Many different forms of gears are used in engineering; different types and gear forms have been designed for specific drives and applications, e.g. worm gears, spur gears, straight-tooth gears, bevel gears and epicyclic gearing.

Robot manipulators are often driven by actuators which are situated some distance away, so the rotary motion of the actuator has to be transmitted by some mechanical means. A typical example of this is the drive to a robot's end-effector. Chains, mechanical links or gears are used to drive bevel gears, which form the wrist joint positioned at the end of the manipulator. Bevel gears are generally used to change the direction of a drive through 90°, but some robot manufacturers have designed drives which use a combination of bevel gears on the wrist that enables all three wrist motions to occur simultaneously about the centre of the wrist joint.

The main problem with using gears is that backlash in the gear train can cause positioning error. If there is no backlash between the mating gears the gears become overloaded, there is excessive noise and overheating. Conversely, too much backlash results in slip between the gears which can lead to servo mechanism instability and hence poor repeatability.

Belt drives

Tooth belts, driven by pulleys, are used to transmit energy over long distances. They have the advantages over gear train drives of greater flexibility and lower cost, and any misalignment between the drive shaft and the output shaft can easily be taken up. As in chain drives, a tensioning device is incorporated in the drive to take up slackness in the belt caused by the belt wearing or stretching.

Harmonic drives

Harmonic drives are used widely on robots to provide a high output torque from robot drive motors and also to reduce the rotational speed with virtually no backlash. Compared with conventional drives, they are lightweight, compact, more reliable and far more efficient. Harmonic drives also provide a unit which has a very low vibration capability at low speeds; this is essential for low-speed and continuous-path applications such as welding and adhesive application.

Figure 2.23 The harmonic drive: the deflection of the flexspine is exaggerated here to illustrate the working principle (*Courtesy Harmonic Drive Ltd*)

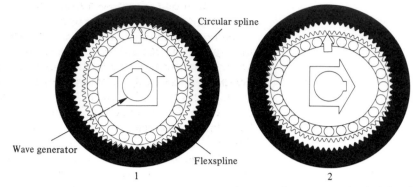

As soon as the wave generator starts to rotate clockwise, the zone of tooth engagement travels with the major elliptical axis.

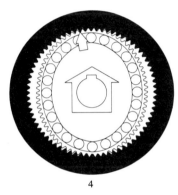

When the wave generator has turned through 180° clockwise the flexspline hase regressed by one tooth relative to the circular spline.

Each turn of the wave generator moves the flexspline two teeth backwards on the circular spline.

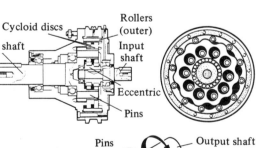

Figure 2.24 The Cyclo speed reducer (*Courtesy Centa Transmissions Ltd*)

The harmonic drive comprises three major components: the wave generator, the flexible steel flexspline and a solid steel circular spline which is rotationally fixed. Figure 2.23 shows these components and explains the principle of operation. The wave generator is a thin-walled ballrace fitted to an elliptical former. When the wave generator is rotated the attached flexspline is elastically deformed, causing teeth to engage with the circular spline across the major axis of the ellipse. The flexspline has two less teeth on its outer circumference, so for each revolution of the wave generator the flexspline engagement point will have moved by this amount in the opposite direction to the input. The flexspline teeth are machined on to the end of the flexspline which forms the output shaft. This is connected to the robot drive linkages, chain sprockets or gears.

Harmonic drives offer high output torques with speed reductions as high as 320:1. The speed reduction can be greater when two or more harmonic drives are coupled together: they can operate at efficiencies as high as 90% – two or three times more efficiently than epicyclic drive units.

Cyclo speed reducers

The Cyclo speed reducer manufactured by Centa Transmissions Ltd employs a unique method of cycloid discs and pins to reduce the speed of, say, an electric robot drive motor. There are essentially four major components in the Cyclo gearbox: a high-speed input shaft with eccentric bearing, cycloid discs, ring gear housing with pins and rollers, and pins and rollers on the slow output shaft. These are shown diagrammatically in Figure 2.24.

The cycloid disc is mounted on the eccentric bearing which rotates at the speed of the input shaft. As the eccentric rotates, it rolls the cycloid disc around the internal circumference of the ring gear housing. The resulting action is similar to that of a disc rolling around the inside of a ring. As the cycloid disc travels in a clockwise path around the ring gear, the discs themselves turn in a counter-clockwise direction around their own axis. Since there is one less cycloidal tooth on the disc than there are rollers in the stationary housing, the cycloid disc rotates in the opposite direction to the input shaft by one tooth pitch for each revolution of the input eccentric and shaft.

Rotation of the cycloid disc is transmitted to the slow-speed output shaft by a number of pins and rollers projecting through holes in the cycloid disc.

The reduction ratio is determined by the number of cycloidal teeth on the disc and the respective number of pins and rollers in the stationary ring housing. Ratios of up to 87:1 are attainable in a single stage but most cycloid gearboxes employ a two-disc system. This serves to double the number of points on the periphery transmitting torque at any one time and by displacing the discs through 180° it ensures perfectly balanced centrifugal forces and smooth operation. Double stage reducers offer ratios of up to 7569:1 with output torques ranging from 10 N m to 6500 N m.

Features of the Cyclo speed reducer include compact size, low weight, low inertia, high torsional stiffness and efficiency as high as 94%.

2.3 End-effectors and end-of-arm tooling

An end-effector is the pickup device – gripper, hand, tool – that is fixed on to the end of the robot manipulator mounting flange. The end-effector is a specially designed piece of equipment that can make use of a variety of techniques and operating principles to accomplish the work that the robot manipulator has to perform.

Considerations in end-effector selection

Various considerations affect the choice or design of end-effector or end-of-arm tooling for its intended task:

- The payload or 'arm capacity' of the robot does not include the weight of the end-of-arm device, which means that extra care must be taken when selecting such devices.
- The design of the end-effector must be flexible enough to allow for multi-task operations – i.e. transfer to other robots – or to cater for families of parts.
- Accessibility is equally important for the end-of-arm device, especially when the application is for machine loading, pick-and-place, assembly work or welding. In addition, the rotational capacity of the robot wrist must allow it enough dexterity to enhance accessibility.
- The size of the working envelope is increased with the fitting of end-of-arm tooling.

Other considerations can be categorized as follows:

The component or object to be manipulated

    Shape
    Size
    Weight
    Position of centre of gravity
    Surface texture
    Location points, if any

Operation of handling

    The positional accuracy needed for the moving of the workpiece or component
    Any joining movements that may be required
    Obstacles in the working area
    Extra movements needed, e.g. remove waste material

Environment

    Sufficient working space
    Temperature of the room/ workshop
    Moisture content of the air
    Excessive/avoidable vibration

Gripping techniques

Various techniques are used to provide the gripping or holding function for a robot end-effector, such as:

Vacuum cups
Electromagnets
Clamps or mechanical grippers
Scoops, ladles or cups
Hands with three or more fingers
Adhesive, or strips of sticky surface

In production processes parts need to be presented to the robot correctly, so that the following considerations will affect the choice of end-of-arm device:

- Parts or items must be held and moved without damage
- Parts must be positioned firmly and rigidly while being operated on
- Hands or grippers must accommodate parts of different sizes
- Self-aligning jaws will ensure that the load stays centred in the jaw
- Jaws must make contact at a minimum of two points to ensure that the part does not rotate while being positioned

Gripper construction

One of the most common methods by which the end-effector controls the component is by use of grippers which may have 'fingers' or 'jaws'. In some applications the gripping action is called *clamping*, which means that the means of release must also be considered.

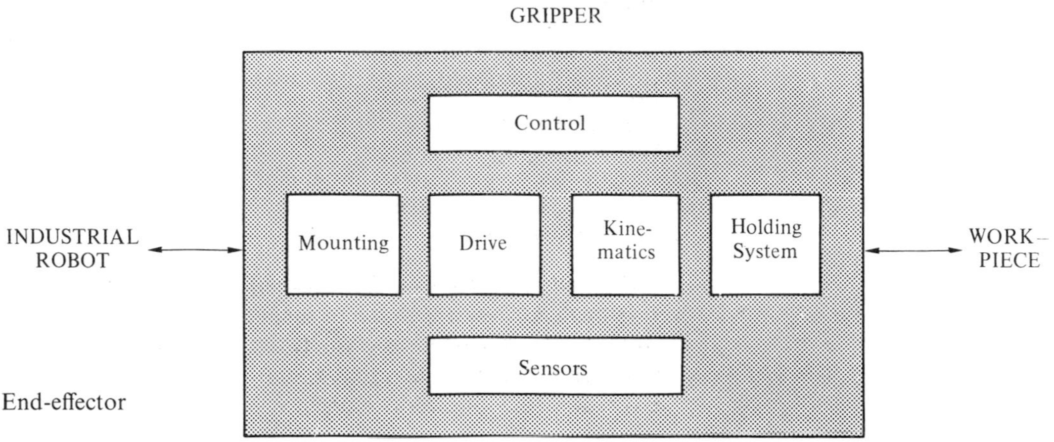

Figure 2.25 End-effector structure

In fact, the structure of the end-effectors consists of different subsections (Figure 2.25). Grippers do not necessarily consist of all the elements shown – mounting plate, drive, kinematics – since in many cases the gripper will only incorporate the mounting plate and a holding system such as a magnet or hook at the end of the robot arm.

The function of the mounting plate is simply to allow the secure connection of the end-effector to the wrist, but it may also be necessary to provide a hand-changing device. Gripper/hand changing is an integral part of a robot program where more than one end-effector is needed to carry out the task.

Gripper drive methods

Power for the movement of the fingers or jaws of grippers and end-effectors may be electrical, pneumatic or hydraulic (Figure 2.26). Table 2.1 shows the application of power sources according to types of gripper drive and drive movement.

Figure 2.26 Gripper drive functions

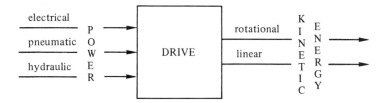

Table 2.1 Application of power sources to drive movement

Gripper drive	Drive movement
Electrical drive	
Stepping motor	rotational
DC motor	rotational
Pneumatic drive	
Pneumatic cylinder	linear
Compressed air motor (high speed)	rotational
Swivel cylinder (low speed, angle of rotation limited)	rotational
Hydraulic drive	
Hydraulic cylinder	linear
Hydraulic motor (angle of rotation unlimited)	rotational
Swivel cylinder (angle of rotation limited)	rotational

Pneumatic devices are the most widely used form of drive for end-effectors because they are cheap to install, the power supply (air) is easy to link to the robot and the system is easily maintained. However, larger systems use hydraulic drives which can cope with heavier loads.

Gripper mechanics

There are many mechanisms available to convert the linear or rotational drive to a linear or finger movement. The drive mechanisms normally used operate either at a constant speed ratio or at a ratio that varies during the finger stroke in such a way that, as the clamping diameter increases, so do the clamping forces that are applied.

Finger movements

Very often simple drive mechanisms with rotational movements are used because they are economical to produce and because of their uncomplicated design (Figure 2.27a, b). These designs are reliable, especially when driven pneumatically, but they have the disadvantage that they may require more computing power when it comes to programming.

When working with parallel faces it is usual for the gripper fingers to be guided in parallel (Figure 2.27c). This ensures that a large area supports the fingers where there are variations of tolerance or differences in the distances between the relative faces.

Care needs to be taken when alignment is needed for irregularly shaped components due to the elasticity of the gripping fingers.

Figure 2.27 Finger movements

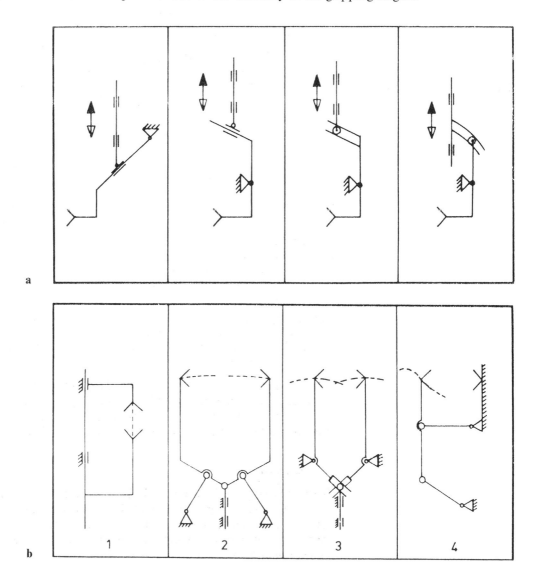

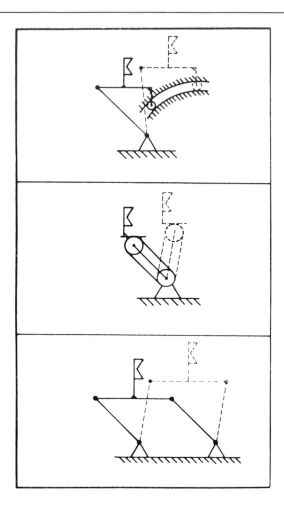

Figure 2.27c

Grippers of this type need to be:

- as light as possible in order to minimize the static and dynamic load imposed by the handling system;
- as small as possible in order to minimize the space required for the grippers in the work area;
- sufficiently rigid for the workpiece to be held accurately in position;
- able to act on the workpiece with sufficient gripping force, without damaging it;
- reliable;
- able to be designed and manufactured relatively inexpensively;
- economical in maintenance expense.

Cylindrical components may be gripped and centred by means of two, three or more point or line systems, as shown in Figure 2.28.

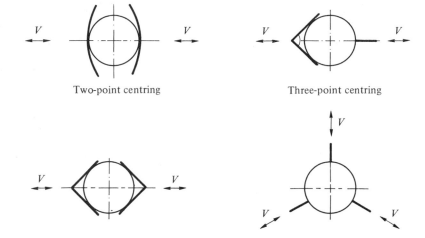

Two-point centring   Three-point centring

Figure 2.28 Methods for
centering cylindrical workpieces

V = closing/opening speed

Four-point centring   Three-point centring

Suction or vacuum grippers

Manufacturers normally supply tables for the calculation of the
vacuum/suction needed to grip and pick up components.

There are three methods of applying vacuum to the suction device:

1 A vacuum pump.
2 The suction device has built-in venturi nozzles which are operated by
compressed air.

Figure 2.29 Vacuum grippers
(*Courtesy IBM Ltd*)

3 Suction cups, which are released by compressed air (Figure 2.29).

Types of end-of-arm tooling

End-effectors and end-of-arm tooling can be classified according to the following areas of application:

Assembly – with hand/tool changing
Machine loading
Welding – spot and seam
Plasma and laser cutting
Machine tool loading

Figures 2.30 to 2.42 show a range of tooling and applications.

Problems

1 State the name given to each axis of an industrial robot with six degrees of freedom.
2 State the six principal configurations of robots in use today and give typical applications.
3 What are the advantages of gantry and overhead-mounted robots over floor-mounted robots?

Figure 2.30 Assembly (*Courtesy Evershed Robotics Ltd*)

4 Define the working envelope of a robot and explain from which point on the robot its dimensions are taken.

5 Sketch the working envelopes of a polar, an articulated-arm and a Cartesian robot.

6 List three ways in which the rotary motion from the drive motor of a robot is converted into linear motion. For each, describe and sketch the way in which this is achieved and any inherent limitations of the system.

7 Describe with the aid of sketches the principle of operation of the harmonic drive system.

8 How can the output ratio of a Cyclo speed reducer be altered? Explain how this device operates and state the advantages of the Cyclo speed reducer compared with conventional gearboxes.

9 What factors are taken into consideration when designing an end-effector for a particular robot task?

10 List six different techniques used to provide a gripping or holding function.

11 Why, when working with parallel faces, are the gripper fingers guided in parallel? Why do these grippers need to be light and rigid?

12 Give three methods of applying a vacuum to suction grippers.

Figure 2.31 Special-purpose compliance gripper for ISO taper cutters (*Courtesy Evershed Robotics Ltd*)

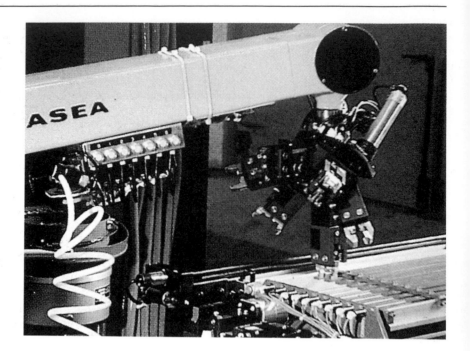

Figure 2.32 Multi-tooling on manipulator (*Courtesy ABB Robotics Ltd*)

Figure 2.33 An A Series robot with hand-changing facility (*Courtesy GMFANUC Robotics (UK) Ltd*)

Figure 2.34 A 600 Fanuc A0 robot with vacuum gripper and auto-hand-changing wrist (*Courtesy Crompton Parkinson Instruments Ltd*)

Figure 2.35 A 600 Fanuc robot moving components by vacuum gripper for drilling (*Courtesy Crompton Parkinson Instruments Ltd*)

Figure 2.36 A 600 Fanuc A0 robot loading a component into a press break (*Courtesy Crompton Parkinson Instruments Ltd*)

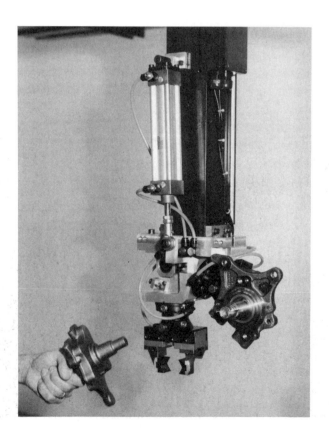

Figure 2.37 Typical double gripper configuration for machine tool loading (*Courtesy Reis Robot Ltd*)

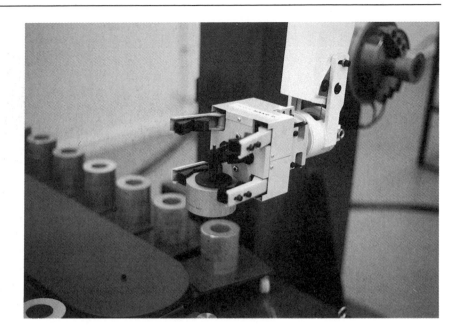

Figure 2.38 Machine loading applications (*Courtesy GMFANUC Robotics (UK) Ltd*)

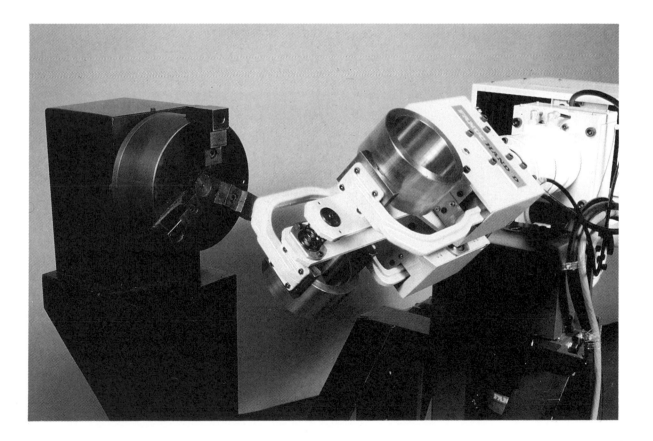

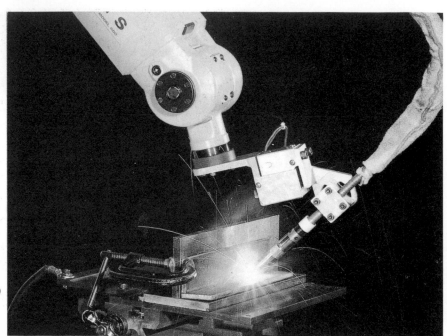

Figure 2.39 600 FANUC S600 MI9, welding (*Courtesy GMFANUC Robotics (UK) Ltd*)

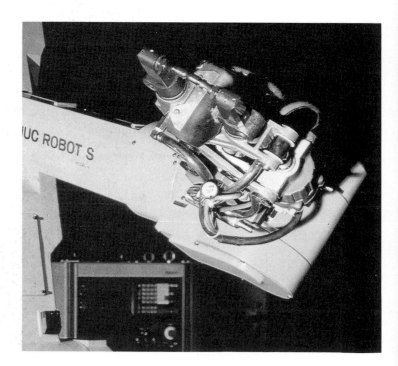

Figure 2.40 600 Fanuc S4 robot with spot-weld gun mounted (*Courtesy GMFANUC Robotics (UK) Ltd*)

Figure 2.41 600 Fanuc robot Model 100, plasma cutting (*Courtesy GMFANUC Robotics (UK) Ltd*)

Figure 2.42 Vacuum gripper (*Courtesy Staubli Unimation Ltd*)

3 Drive systems

3.1 Hydraulic drives

In early robot systems the hydraulic drive and the pneumatic drive were the accepted means of powering the movement of robot arms within their working envelopes. In the natural progression of research into drive methods, the use of computer power combined with electric drive systems was developed. Pneumatically and hydraulically driven robots came to be produced in smaller numbers and used in fewer applications. However, hydraulic drives have proved more suitable for certain purposes, notably for lifting heavy weights, e.g. lifting beer barrels in a palletizing routine.

Various devices can be used to drive the robot arm hydraulically, one of these being the basic piston or transfer valve drive unit. In using such devices, it is necessary to take into consideration:

Flow rate
Dimensions of the piston and valve
Coefficient of friction
Orifice flow
Mass
Flow acceleration requirements

Hydraulic piston transfer valve

The basic units and the equations that relate them are shown in simplified form in Figure 3.1, which shows that the total force exerted by the piston will be equal to the differential pressure, shown as $P_1 - P_2$, and that this pressure will in turn be governed by the flow rate of oil through the transfer valve into the piston.

Note that the rate of change of the shaft position, $Y(t)$, turns out to be a constant times the transfer valve shaft displacement, $X(t)$.

Hydraulic circuit incorporating control amplifier

Taking this concept a little further, Figure 3.2 shows a slightly more elaborate layout using a control amplifier with its many inputs, the solenoid, and a load reaction. The actual transfer of power, from which the arm position sensor feeds information back to the control amplifier, is equal to the expected transfer force minus the load reaction.

Another simplified method which can be used to develop the equation of motion for the hydraulic piston valve and load is shown in Figure 3.3. However, we must realize that nature does try to balance the forces

Figure 3.1 Hydraulic piston transfer valve system

$$P_L = P_1 - P_2$$
$$Y(t) = \text{constant} \times X(t)$$

$$Q = CA \sqrt{\left(2g\frac{\Delta P}{W}\right)}$$

where:

A = orifice area
g = acceleration due to gravity
ΔP = pressure drop across inlet
Q = flow rate of oil
W = specific gravity of oil
C = orifice coefficient

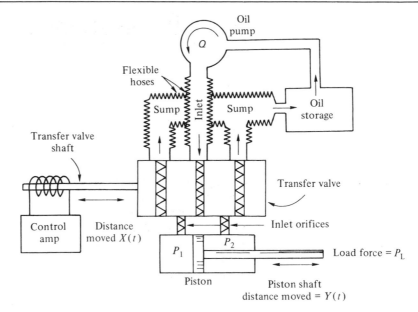

represented. The piston force tends to make the load accelerate and move the mass of the load, whereas the friction in the load system, or robot arm, tends to resist motion. There must be a force greater than the retarding forces if the system output is to be moved and there is to be an acceleration of the load.

The load might consist of the members of the robot arm system, or the arm plus whatever the arm is lifting or moving. In any event, the desired result is that the arm and load move smoothly and accurately without vibration, oscillation, overshoot or any other problem.

Remember that in a hydraulic system the forces can be quite considerable since the force exerted on the piston shaft is the product of the pressure of the oil times the area of the piston.

Some of the favourable aspects of the hydraulic system are low weight per unit power, low inertia of output member, high speed of operation and large force development. Its limitations include the requirement for an auxiliary power system, possible leakage problems and minimal control.

Figure 3.2 Hydraulic power actuating system specifying transfer functions

$$P_t = F_t - L$$

where:

P_t = actual transfer of power
F_t = expected transfer force
L = load reaction

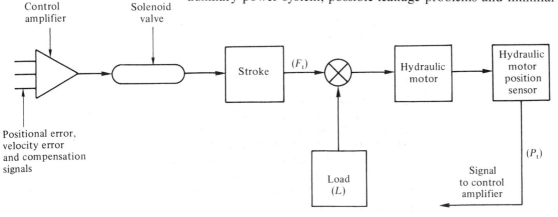

Figure 3.3 Rate of hydraulic force
transfer = (mass × acceleration) + (friction × velocity). Therefore, with respect to the distance moved, S, in time, t,

$$AP(t) = Ma(S) + Fv(S)$$

where:

P = pressure
A = area
M = mass of the load
F = friction force
a = acceleration
v = velocity

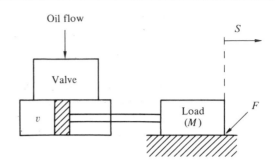

The early hydraulic robot

One of the most popular ranges of hydraulically driven robots was produced by Cincinnati Milacron in the USA. The two robots were models T³-566 and T³-568; the latter was for heavy duty application. Both robots had six degrees of freedom, being direct electro-hydraulic servo drive systems (see Figure 2.21).

Each axis had its own positional feedback device, consisting of a resolver and a tachometer to ensure accurate arm repeatability. The following specifications for the T³-566 and T³-586 show how adaptable these robots could be for many industrial applications.

Load capacity

Load 250 mm from tool mounting plate, T³-566 Robot	45 kg
Load 250 mm from tool mounting plate, T³-586 Robot	100 kg

Number of axes, control type

Number of servoed axes, hydraulically driven	6
Control type	Controlled path at tool centre point

Positioning repeatability

Repeatability to any programmed point	± 1.25 mm

Jointed arm motions, range velocity

Maximum horizontal sweep	240°
Maximum horizontal reach to tool mounting plate, T³-566 Robot	2460 mm
Maximum horizontal reach to tool mounting plate, T³-586 Robot	2590 mm
Minimum to maximum vertical reach	0 to 3900 mm
Maximum working volume	28 m³
Maximum velocity of tool centre point (TCP), T³-566 Robot	1270 mm/s
Maximum velocity of tool centre point (TCP), T³-586 Robot	890 mm/s
Pitch	180°
Yaw	172°
Roll	270°

Memory capacity
Number of points which may be stored   1750

Floor space and approximate net weight
Robot $0.8\,m^2$ 2270 kg
Hydraulic and electric power unit $1.5\,m^2$ 540 kg
Acramatic control $0.8\,m^2$ 365 kg
Power requirements   460 volts, 3 phase, 60 Hz
Environmental temperature $5°$ to $50°C$

Hydraulic fluid considerations

Although there are systems that are powered by water, they present problems of oxidation, growth of mould and poor lubrication. The most useful hydraulic fluids are in fact mineral oils with viscosities of 20–30 centistokes at the operating temperature. These fluids are also excellent carriers of heat dissipation to the exterior and therefore contribute to safer running conditions and temperatures.

Although in some robotics systems pressures can reach several hundred bar, the upper limit is usually about 100 bar (10 MPa), which is compatible with the use of light and flexible tubes. The greater mass and relative rigidity of feed tubing actually detract from the advantages of hydraulic systems compared with pneumatics, also the gain in power-to-weight ratio becomes negligible. Therefore the main problems to be resolved are:

1 Transport of oil under pressure to the areas in which it can be used, when these are distributed around a polyarticulated mechanical system of highly variable geometry.
2 Leakage from connecting pipes.
3 Adequate fluid filtration – particles of dimensions greater than 5 μm block servo valves, erode flow controllers in the distributors and cause deterioration in the state of internal surfaces of pistons as well as increasing the viscosity and the apparent compressibility of the oil.

Hydraulic fluid must be carefully conditioned before it is allowed into a hydraulic circuit. It must first of all be filtered of any dirt and foreign particles. Such particles could prevent the correct operation of precision components with unsafe consequences, or contribute to premature component wear or damage. All air within the system must be similarly removed. The presence of air can cause compressibility of the fluid and also lead to a condition called *cavitation*, which can cause extreme wear and damage to circuit elements if allowed to persist. All leaking joints, as well as reducing the efficiency of the circuit, are also potential entry points for air and dirt. The efficient sealing of hydraulic cylinders is a constant problem and the cause of much associated long-term maintenance. Cooling of the fluid may also need to be carried out to maintain an acceptable working temperature of the hydraulic fluid.

Hydraulic accumulation

An important consideration is that there is no distribution network for pressurized oil to workshops as there is for electricity and compressed air. Therefore for hydraulic systems it is necessary to establish a 'hydraulic centre' close to all machines which can provide a regulated, pressurized supply by means of an electric motor and a hydraulic pump. The size of the fluid reservoir must take into account the heat to be dissipated. Often accumulators are used at the output of the central hydraulic unit, or around devices with high instantaneous consumption.

The accumulator is a fluid storage/heat-dissipating container made up of a reservoir divided into two compartments, separated by a folded membrane which can sweep through the whole volume of the reservoir. One compartment is filled with air or with an inert gas (e.g. nitrogen), at the minimum operating pressure for the hydraulic circuit, and is compressed when the hydraulic fluid is introduced into the complementary compartment. It acts as a spring, reducing the battering-ram effect when circuits are opened or closed by providing or absorbing flow. These hydraulic systems function at nearly constant average pressure.

Hydraulic activators

There are three main types of hydraulic linear activator, which are very similar to pneumatic pistons but are smaller and lighter (for a system of equal power):

Double-action pistons
Differential pistons
Single-action pistons

Double-action piston

The double-action piston activator (Figure 3.4) has two chambers, being supplied at pressures p_1 and p_2. Because the output shaft is in the second

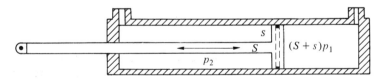

Figure 3.4 Double-action piston

chamber, the effective cross-sectional areas of the pistons $(S+s)$ are different and the force (f) developed is not a symmetrical function of the pressures that exist in the two parts.

The equation derived from this statement is shown as

$$f = (S+s)p_1 - Sp_2$$

However, this drawback can be overcome by using a piston with a double shaft, as shown in Figure 3.5. This design of shaft will enable the equation to

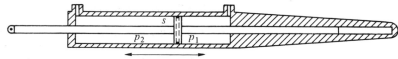

Figure 3.5 Double-action piston with double shaft

be written as

$$f = s(p_1 - p_2)$$

and, if p_a and p_b are the supply and return pressures respectively, then the maximum force is

$$f_m = s(p_a - p_b)$$

where s is the active surface (the area of the piston reduced by the area of the shaft).

Differential piston

In the differential piston (Figure 3.6), the shaft section is equal to half the total area of the piston, i.e. $s = S$. The chamber through which the shaft

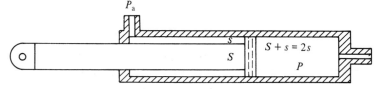

Figure 3.6 Differential piston

passes is permanently connected to a high pressure source p_a, whereas the other chamber is subjected to a variable pressure, p. Therefore the force developed by this type of piston is

$$f = S(2p - p_a)$$

with maximum force f_m equivalent to that of a symmetrical double-action piston but half of that of a double-action piston with a single shaft. A differential piston requires only a three-track gate, whereas a double-action piston requires a four-track gate of a more complicated structure.

Single-action piston

The single-action piston (Figure 3.7) is a simple construction with a single input port. It has an active chamber with variable pressure p and the other is at low pressure p_b. The force developed is in a single direction:

$$0 \leqslant f = S(p - p_b) \leqslant S(p_a - p_b)$$

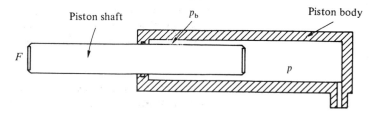

Figure 3.7 Single-action piston

Therefore the piston return must be effected by an external force, which limits the application of this particular design.

The three types of piston described can develop large forces and long strokes. The asymmetrical double-action piston is more limited with regard to stroke than the others, because its shafts must be thin in order to minimize the imbalance of stress and the shaft therefore has less resistance to buckling. However, for long strokes the differential piston is the best choice and allows the longest reach. The ratio of the stroke to the diameter of the piston reaches 50.

3.2 Fluid power systems

Fluid power systems involve a number of conversion stages in order finally to achieve axis motion. Initially, the hydraulic oil has to be pressurized, generally by an electric motor driving a hydraulic pump which delivers the hydraulic oil under pressure. Since the electric motor is the start of the conversion process, it is termed the *prime mover*. (Earlier hydraulic systems often used internal combustion engines as the prime mover.) The fluid power so generated is then converted into mechanical power via hydraulic cylinders or rotary actuators. Losses in conversion are small, usually manifesting themselves as heat, vibration and noise. Hydraulic systems require a closed circuit, and exhausted fluid must be directed back to the pump (by return lines into a pump tank) for recirculation.

Figure 3.8 Robot hydraulic system

Figure 3.8 explains how the components described below fit in to the robot hydraulic system.

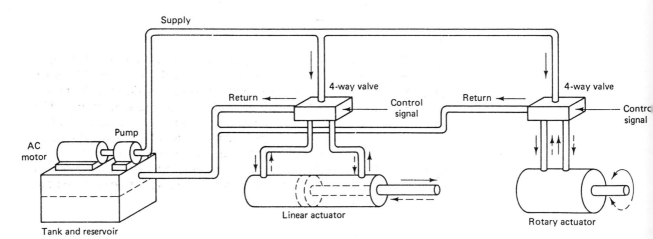

Hydraulic pumps

Hydraulic pumps may deliver fixed quantities of oil per motor revolution. They are termed *fixed displacement pumps*. These pumps are acceptable when demand for pressurized fluid is relatively constant. As robot axes operate, they can often vary the demand on the hydraulic pump. To accommodate this, *variable displacement pumps* are employed. These

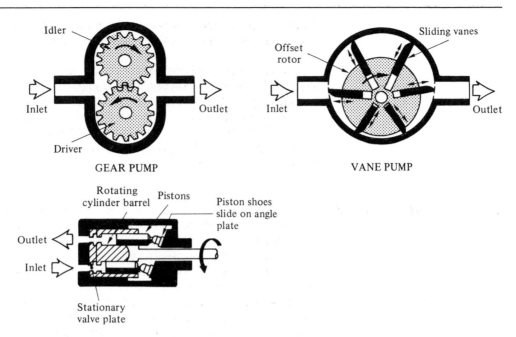

GEAR PUMP

VANE PUMP

PISTON PUMP

Figure 3.9 Hydraulic pumps

pumps are capable of varying the volume of fluid delivered (at constant pressure) according to demand. The principles of operation of the common hydraulic pump types are illustrated in Figure 3.9.

Actuators

Hydraulic fluid systems may drive rotary vane type actuators called *rotary actuators*, or be used to extend or retract pistons within cylinders. The latter are termed *linear actuators* and are employed on prismatic joints. Note that engineers refer to hydraulic cylinders rather than hydraulic pistons. The principles of rotary and linear actuators are illustrated in Figure 3.10.

The orifices in hydraulic components that admit hydraulic fluid and allow it to be exhausted are termed *ports*. The ports may act as inputs or outputs for the hydraulic fluid, depending in which direction the fluid is routed. This allows both rotary and linear actuators to be powered in both forward and reverse directions.

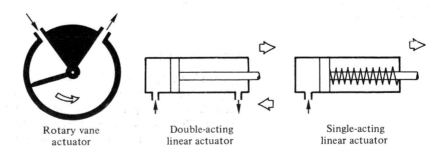

Figure 3.10 Rotary and linear actuators

Rotary vane actuator

Double-acting linear actuator

Single-acting linear actuator

The rotary actuator comprises a shaft joined on to a rotary vane that separates two ported chambers. As hydraulic fluid is displaced from one chamber (by pressurized fluid entering the other), the vane rotates and drives an output shaft.

The linear actuator works on the same principle. Some cylinders can only be extended under fluid power. Retraction of the piston is accomplished by a compression-type return spring accommodated within the cylinder body. These are termed *single-acting cylinders*. For more precise control, and power availability in both forward and reverse directions, there are *double-acting cylinders*. These cylinder types (Figure 3.10) are available in various standard sizes, in a number of designs and offering a number of different mounting arrangements.

Bird Johnson HYD-RO-AC

This rotary actuator converts hydraulic pressure into reciprocating output shaft movement through an arc up to 280° total travel. The working mechanism of the actuator is enclosed and sealed to prevent the entrance of foreign material.

In the single-vane actuator (Figure 3.11), when high-pressure fluid enters port 1 which is connected to chamber A, it causes chamber A to increase in volume. Chamber B reduces in volume. The fluid flow caused by high fluid pressure acting on one side of the wingshaft vane, with low pressure on the other side, rotates the wingshaft in a counterclockwise direction. When high pressure fluid is applied at port 2, and port 1 is connected to discharge pressure, rotation in the opposite direction takes place.

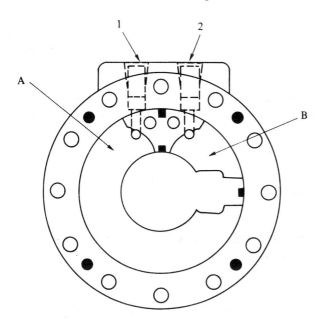

Figure 3.11 Single-vane actuator

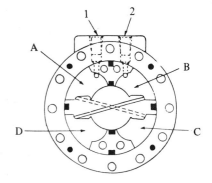

Figure 3.12 Double-vane actuator

In the double-vane actuator (Figure 3.12), when high pressure fluid enters port 1, which is connected to chamber A, it causes chamber A and chamber C, which is connected by internal porting, to increase in volume. The other pair of chambers (B and D), which are connected to the discharge port, will reduce in volume. The fluid flow caused by high fluid pressure acting on one side of the wingshaft vanes, with low pressure on the other side, rotates the wingshaft in a counterclockwise direction. When high pressure fluid is applied at port 2 and port 1 is connected to discharge pressure, rotation in the opposite direction takes place.

The torque developed by the actuator is proportional to the area of the wingshaft vane and the hydraulic fluid pressure differential. Speed of rotation is dependent on the flow and pressure capacities of the hydraulic system.

External stops are used to limit angular travel as the actuator abutments are not designed as mechanical stops.

Shaft attachment and mid-travel position: the actuator uses a multiple-tooth involute spline which is self-centering, ensuring equal distribution of the torque load over all serrations. The mid-position of travel is obtainable by positioning the missing tooth area of the spline 180° opposite the centreline of the hydraulic connecting ports for single-vaned units, and 90° clockwise from the centreline between the two ports for double-vaned units (Figure 3.13).

Special tools are required for reassembly.

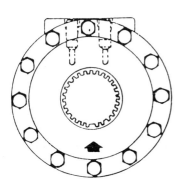

Figure 3.13 a Single-vane mid position

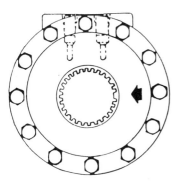

b Double-vane mid position

Specifications

Operating pressure:

Maximum rated pressure	207 bar (3000 psi)
Operating pressure range	6.9 to 207 bar (100 to 3000 psi)

Output torque:

Input pressure from 6.9 to max. 207 bar (100 to 3000 psi) provides torque from 5.6 to 83.659 Nm (50 to 741 000 inch pounds) with efficiency of at least 90%.

Angular travel:

Total shaft travel for standard single vane units is 280° ± 5°.
Total shaft travel for standard double vane units is 100° ± 5°.

Angular velocity:

Angular velocity can be controlled by metering the amount of flow of fluid into or out of the actuator parts.

Hydraulic valves

The control of pressure, flow and direction of hydraulic fluid within a hydraulic circuit is accomplished by discrete hydraulic valves. The most common type is the *spool valve*. A precision-ground shaft (the spool) moves horizontally within the valve body. In so doing, it either covers or uncovers inlet or outlet ports, thus allowing the passage of fluid. Spool valves may be operated in a number of ways: manually, or by pneumatic, electrical or small hydraulic signals. The latter are termed *pilot operated valves*, since only a relatively small signal (the pilot signal) is required to actuate them. Control valves differ in the number of ports they have, the number of ways in which they allow fluid to pass, and their method of actuation. The principle of operation of a typical spool valve is illustrated in Figure 3.14.

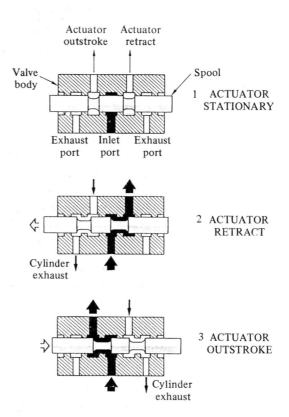

Figure 3.14 Spool valve

Hydraulic circuits are connected via rigid (steel) pipes or by flexible (reinforced) hoses.

MOOG electrohydraulic servo valves

These are two-stage flapper valves consisting of a polarized electric torque motor and two stages of hydraulic power amplification (Figure 3.15). The torque motor armature extends into the air gaps of the magnetic flux circuit and is supported in this position by a flexure tube member. The flexure tube also acts as a seal between the electromagnetic and hydraulic sections of the valve. Two torque motor coil surround the armature, one on each side of the flexure tube.

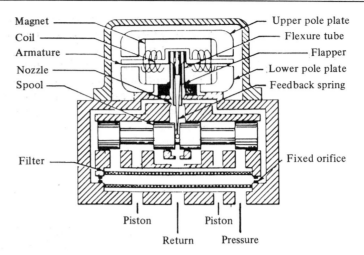

Figure 3.15 Servo valve

The flapper of the first-stage hydraulic amplifier is rigidly attached to the midpoint of the armature. The flapper extends through the flexure tube and passes between two nozzles, creating two variable orifices between the nozzle tips and the flapper. The pressure controlled by the flapper-nozzle variable orifice is fed to the end areas of the second-stage spool.

The second stage is a conventional four-way spool design in which output flow from the valve, at a fixed valve pressure drop, is proportional to spool displacement from the null position. A cantilever feedback spring is fixed to the flapper and engages a slot at the centre of the spool. Displacement of the spool deflects the feedback spring which creates a force on the armature/flapper assembly.

An input signal induces a magnetic field in the armature and causes a deflection of the armature and flapper. This assembly pivots about the flexure tube, increasing the size of one nozzle orifice and decreasing the size of the other. This action creates a differential pressure between one end of the spool and the other, and results in spool displacement. Spool displacement deflects the feedback spring, generating an opposing torque to the original input signal torque. Spool movement continues until the feedback wire torque enables the input signal torque.

3.3 Pneumatic drives

Pneumatic fluid power systems use compressed air as the main transmission medium. Many of the principles involved in hydraulic system operations also apply to pneumatic system operation. Since air is compressible, precise control of speed and position is more difficult to achieve, and power application is less than that which is achievable in hydraulic systems. However, the compressibility of air can also be used to advantage when absorbing shock loads, preventing damage due to overloading, and also provides an element of compliance or 'give' which may be exploited in many practical applications. The advantages of pneumatic systems include the following:

- Air is plentiful and compressed air is readily available in most factories.
- Compressed air can be stored and conveyed easily over long distances.
- Compressed air need not be returned to a sump tank: it can be vented to atmosphere after it has performed its useful work (but exhausting can be an irritatingly noisy process).
- Compressed air is clean, explosion-proof and insensitive to temperature fluctuations.
- Operations can be fast and speeds and forces can be infinitely adjusted between their operation limits.
- Digital and logic switching can be performed by pneumatic fluid logic elements.
- Pneumatic elements are simple and reliable in construction and operation and are relatively cheap.
- Pneumatic cylinders comprising proximity sensors for sensing the positions of the pistons allow easy integration of pneumatic systems with computer-aided service and control.

As with hydraulic systems, a number of conversions have to be carried out, which pass through several stages. Compressors, driven by a prime mover, draw in and compress air from the atmosphere. They may be either piston compressors or turbine compressors, both types being available in a number of designs. Piston compressors operate by reducing the volume of air and increasing its pressure. Turbine compressors operate by drawing in air, compressing it by mass acceleration and converting kinetic energy into pressure energy.

Compressors are specified according to their delivery volume and their delivery pressure. Air delivered by the compressor can be fed directly into the circuit (after conditioning), or accumulated in a pressure vessel called a receiver. The receiver, as well as acting as a storage container, provides a reservoir for sudden demands in air pressure.

Before compressed air is admitted into a circuit it must first be conditioned, to ensure reliability of circuit operation. Conditioning, carried out by special-purpose service units, performs three principal functions:

1 All dirt and foreign particles are filtered out. This increases the system reliability by lessening wear and damage. It also reduces the incidence of jamming or blocking up of delicate control devices, and the misoperation of circuit components.
2 The air is dried to remove absorbed moisture which could otherwise condense out within the circuit components.
3 The air is enriched with a very fine oil mist which provides lubrication for the various system components.

Compressed air can be applied to both linear and rotary actuators. Linear actuators in the form of air cylinders are quite similar in construction and principle of operation to their hydraulic counterparts. Rotary actuators, which are commonly termed 'air motors', consist of a rotary vane driven by the applied air pressure.

Pistons

Linear pistons

The dual-action linear piston (Figure 3.16) is generally on–off controlled, and the speed of the movement is not precisely governed. At each port an adjustable restriction, known as a 'flow reducer', is placed. This is made unidirectional by placing a one-way valve in parallel, to allow free passage of air during admission.

However, to overcome friction and to avoid any juddering movement, it is important that the maximum theoretical pressure in the section of pipe should be greater than any opposing pressure (generally by a factor of 2). Asymmetrical forces are developed in the pulling and pushing movements of the actuator and the size of pistons intended for long-term operation should take into account the tendency of the shaft to flex and buckle under load.

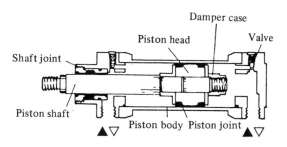

Figure 3.16 Dual-action linear piston

When a large mass is to be moved, the shock due to the insertion of the mass at the end of the stroke can be reduced by arranging for a progressive deceleration over the final distance of a few centimetres (without rebound) by means of a damper system, which can be used to absorb the kinetic energy of the load and dissipate it as heat.

If a pneumatic piston is to be used over its full stroke, it is useful to employ a damped piston, i.e. one with pneumatic dampers at each end. The piston includes two end-sections of reduced diameter which constitute a damping chamber for each end of the stroke, with an adjustable screw to control the escape flow.

To control the stroke distance in this kind of piston, mechanically adjustable limits are used. Two are required for a single-stroke piston and more for multiple-stroke systems. Attached to each limit should be an external damper with its own piston. This is usually an adjustable hydraulic damper, operating in a closed fluid circuit.

Pistons are usually made from a copper alloy or steel, and the body is made of steel. However, some pistons are made with cylinders of polyester resin reinforced with glass fibre. This particular combination of materials allows magnetic-checking positional-correction switches to be placed directly into the cylinder and activated by movement of the piston arm.

Rotary pistons

If the amplitude of rotation is to be limited (e.g. in a manipulator wrist), the actuator most frequently used is of the type shown in Figure 3.17. The shaft carries a piston-operated rack that drives a pinion on the output axis. There are, however, other systems that transform pressure to a rotational movement (e.g. screw-nut systems, cams).

Special systems

The following are systems that have been found to be useful in automation:

- Multi-stage pistons that allow several positions to be achieved
- Impact or percussion pistons for riveting or marking operations
- Cable pistons without shafts for long strokes
- Curved pistons without shafts for lateral outputs
- Membrane or bellows pistons for long strokes
- Simple pistons or 'artificial muscles' which swell and shorten as the internal pressure increases

Motors

A motor is primarily a speed generator that allows moving parts to be driven over a distance which is not defined. In pneumatics two types of motor are encountered, volumetric motors and turbines.

Volumetric motors are largely the same as hydraulic motors and consist of:

Flapper motors
Geared motors
Piston and membrane motors

The motor velocity is directly related to flow rate, whereas torque is determined by the pressure applied.

The turbine is a widely used actuator of unlimited rotation. It consists of a pneumodynamic system that can develop torque and revolve at very high speeds of up to 300 000 rpm. Turbines use a gas containing a suspension of lubricating oil. This can cause considerable corrosion if any leaks occur. They are used mainly in conjunction with specialized end-effectors, such as drills for tightening screws.

Components used in pneumatic control

The supply to pneumatic motors is generally binary, modulated by distributors or switches, and communication direction is achieved either by pneumatic movement (via relay membranes) or by electrical movement (movement caused by the use of an electromagnet with a movable core or a rotary arm) of the mobile parts. Distributors can be divided into two main

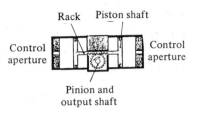

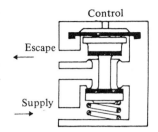

Figure 3.17 Rotary piston

Figure 3.18 Valve distributor

types, depending on whether the moving part is a valve or a spool (linear movement).

The valve distributor

Figure 3.18 shows a distributor with three apertures and two positions. The valve is moved by a membrane activated by a control-pressure command and returned by a spring, making it a main stable distributor.

The valve is moved from one position to the other, opening and closing the air passages in these sections as it does so. This technique is especially useful for distributors with two or three apertures. It operates without friction and therefore does not require lubrication, but the operation of the valve is sensitive to pressure and this limits the use of more stable variants.

The spool distributor

A mobile spool moves across ports, varying the amount that they are opened and hence controlling the movement of air between the ports (Figure 3.19). In bi-stable distributors the spool is moved by the difference

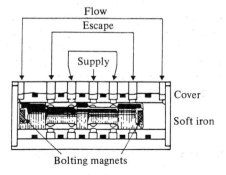

Figure 3.19 Spool distributor

in pressure between the end ports, and maintains pressure with no pressure difference. This is ensured by the spool being balanced relative to the pressure of the control and input and output ports.

3.4 Electric motor drives

An electric motor is a transducer that converts electrical energy into mechanical energy by means of a DC or AC power supply. These motors are used as actuators in many industrial robots for many reasons, the most important being that they generate very high torques for their size and are easily controlled by a microprocessor.

DC electric motors

The DC motor operates on the principle of an electrical current flowing through a wire-wound armature or rotor which is positioned inside a magnetic field. This magnetic field is developed either from permanent magnets or by means of an electromagnet called a *stator*. Connected to the armature is the commutator through which the supply current flows via carbon brushes.

Figure 3.20 illustrates the DC motor components and the principle of rotation. A DC current is supplied to the brushes which is passed to the commutator and so to the wire-wound armature which creates a magnetic field. The polarity of the stator magnetic field is the same as that of the armature. As both magnetic fields are of the same polarity they repel each other and the armature turns in a clockwise direction. When the armature has rotated through 90° it comes under the influence of the opposite polarity from the permanent magnet; since opposite poles attract, the armature is pulled in the clockwise direction, i.e. rotated through a further 90°. When in this position, the magnetic field in the commutator is changed so the stator now repels the armature, thus causing it to rotate a further 90°.

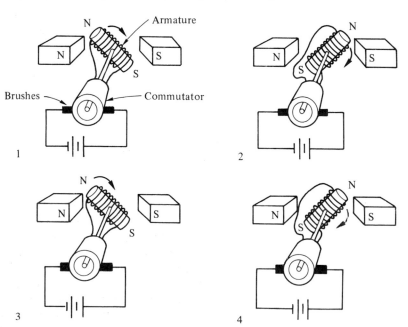

Figure 3.20 Operating principle of a DC motor

The armature is then attracted to the unlike polarity so it continues through a further 90°, and on reaching this position the polarity in the armature is changed so the two fields again repel each other. The rotation of the armature obviously continues until the power supply is terminated.

To overcome the problem of brush wear and consequent arcing in the motor, DC brushless motors are now used. These use solid state electronics to switch the power supply in the armature, so making the motors more reliable and safe in hazardous environments.

Various types of DC motor are used for robot manipulators; all operate on the principle described, the differences between them consisting of the ways in which the stator winding is designed and the types of material used in their construction. For robot wrists, series-wound motors are used because their speed decreases with the application of load: with light loads the motor accelerates rapidly, so enabling the end-effector to reach the desired position at high speed. Shunt-wound motors offer the advantage of producing a constant speed irrespective of the load applied to the motor; this means that a constant torque is developed from the drive. This type of design enables changes of speed to be implemented, i.e. the execution of different programmed feedrates, so making them more flexible than the series-wound motor.

AC electric motors

In many of the newer robots an AC motor replaces the DC motor as the actuator on the manipulator axes. The reason for this is that the AC induction motor can produce the same power output as a DC motor but with a higher torque and a reduction in size. Added to this, the AC motor does not require brushes or commutator, is self-cooling and is completely enclosed, thus eliminating maintenance requirements.

The operation of an AC induction motor can be simply explained. An AC voltage is applied to windings on a stationary stator, producing a magnetic field around it. This induces a current to flow in the rotor conductor which is situated in the stator field, and this in turn causes a magnetic field of opposite polarity around the rotor: as the two fields repel each other, the rotor turns and hence an output torque is developed.

Stepper motors

Stepper motors are not generally used for driving the axis movements of industrial robots but they are used for various robot applications where low torques are required and the control system is operating as an open loop. Unlike the DC motor, the rotor has teeth or steps and rotates in magnetic fields generated by multiple permanent magnets in the stator.

Figure 3.21 shows the principle of operation of a permanent-magnet stepper motor. The polarity of the magnets in the stator poles is changed when electrical currents are pulsed into the stator windings. When one stator pole is turned on, the stator pole next to it is turned off. This turning on and off of stator poles, and the resulting change in polarity of each stator pole from north to south, causes the rotor to be attracted from one pole to the next thus causing the rotor to rotate. Each step is precise and the number of steps, or angular increments, for one complete revolution is dependent on the number of stator windings; usually the range is between 1.8° and 30°.

The disadvantages of stepper motor drives are that they are not suitable for heavy loads and they do not give a smooth drive since there is acceleration and deceleration between steps. Also, special ramping circuits have to be incorporated in their design to give a very high pulse rate on acceleration, so as to generate the required torque to move the load, with the reverse requirement on deceleration. They are, however, relatively inexpensive and are ideal for moving light loads to precise positions using open-loop control.

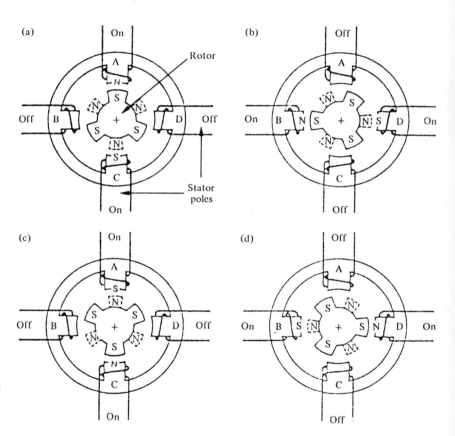

Figure 3.21 Operating principle of a permanent magnet stepper motor

Stepper motors are used in many robot applications for driving peripheral equipment such as conveyors. They are also used on special-purpose end-effectors, e.g. the windscreen insertion end-effector on the Montego line at Austin Rover.

Problems

1 What dimensions and values are critical to the use of a hydraulic drive unit?
2 Discuss the performance specifications of the early American hydraulic robot.
3 Name some of the problems caused by using water to power a hydraulic drive system.
4 Draw a single-action piston and explain its principle of operation.
5 Explain the problem of cavitation in a hydraulic system.
6 Explain the operation of the double-vane actuator.
7 Describe the function of the two-stage flapper valve.
8 List the advantages of pneumatic drives.
9 What devices can be used to convert fluid pressure into rotational movement?
10 For what reasons do robot manufacturers use electric motor drives in preference to hydraulic drives?
11 Explain why AC electric motors are used to drive manipulators rather than DC drives or stepper motors.
12 Draw and discuss the components of a DC electric motor.
13 Describe with the aid of diagrams the operating principle of a stepper motor and explain (a) why no feedback loop is necessary and (b) why a ramping circuit is required.

4 Robot sensors

4.1 Sensing requirements

In order to operate efficiently, robots must be fitted with sensors to gather and evaluate information about the environment in which they are sited and to recognize the component or process on which they are working. Humans automatically react if a component is too heavy or bulky to lift, or too hot to handle; we immediately adjust the force of our grip when we recognize that a fragile part requires a delicate grasp in moving it from one place to another. Added to this, we have vision to detect colour, identify particular parts and appreciate their orientation. We also have the inherent ability to carry out visual inspection of all manner of complex forms.

One day robots may have to some degree the senses that we possess of touch, smell, sight and hearing; at present tactile and visual sensing are the two areas which are attracting the most interest in research and development.

Sensors for robots can be considered to fall into two main categories, contact and non-contact sensors. They can also be classified, according to their complexity, into 'simple' and 'intelligent' sensors as shown in Figure 4.1.

The purpose of incorporating sensors in a robot cell includes:

- The protection of the end-of-arm tooling, peripheral devices and workpiece from damage which could be caused by collision, by entry into a hazardous environment or by the forcing of parts together in an assembly operation.
- The reduction of cost by identifying randomly orientated parts and acting upon them appropriately, so eliminating the need for bowl feeding devices and complex sorting devices.
- The gathering of real-time information such as part orientation, size, shape, surface finish, colour, position and specific mass so that the robot task can be carried out in the minimum cycle time.

4.2 Tactile (contact) sensors

Tactile sensing may be defined as the ability of a sensor to determine the shape, size, weight or even the surface texture of a part by touching it. Thus this type of sensor is incorporated in the design of the gripper or part of the end-effector which is making contact with the workpiece. It follows that tactile sensors provide information on the distribution of forces in the wrist and in the joints of the robot as well as in the objects that are to be handled. The detection of forces and of their distribution is of particular importance in such applications as fettling and grinding and also in assembly operations.

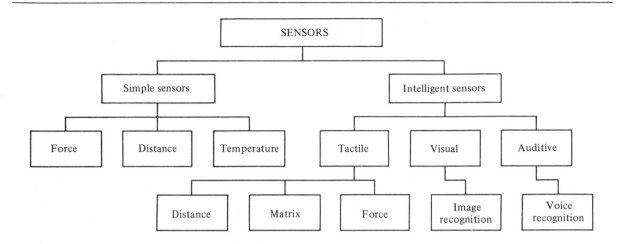

Figure 4.1 Robot sensors classified according to complexity (*Courtesy KUKA Welding Systems and Robots Ltd*)

The size and weight of these sensors can be an important consideration since they may add considerable weight and bulk to the end-effector. This results not only in an increase in the inertial effect of the robot arm but also the physical size of the sensor might restrict access to the component or to the parts feeding area.

There are many types of tactile sensor used in robotic applications, the most common being microswitches, piezo-electric devices, strain gauges and pressure-sensitive materials.

Microswitches

The simplest type of tactile sensor is the microswitch operated by a lever, roller or probe which makes or breaks a circuit when actuated by a tool, workpiece carrier or, perhaps, the arrival of a part to be picked up from a conveyor or bowl feeder. The microswitch sends a closed-switch signal to the controller, indicating that the part is present. The gripper can incorporate a microswitch to tell the controller that the part has been picked up, or that it has not been located. Also it can be used to indicate that the part has been crushed or has fallen out of the gripper.

It is important to remember that these switches are fragile: care must be taken when moving parts or the robot arm into contact with them.

Piezo-electric sensors

These sensors use the piezo-electric effect to quantify the amount of force that is being exerted by the gripper. Quartz or lithium niobate crystals in the form of wafers are attached to the gripper faces. Upon contact with an object, the wafers are stretched or compressed, and this deformation causes the material to become electrically polarized. The small e.m.f. generated is amplified and sent to the robot controller.

These devices can be used for the measurement of force, pressure and surface texture. When arranged in the form of a matrix they can be used to sense the surface contour of an object.

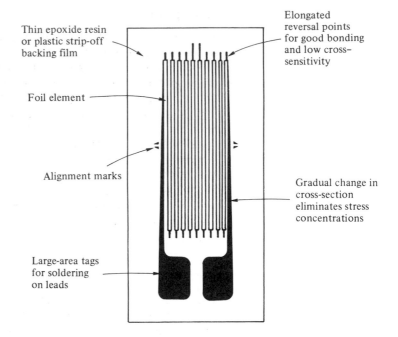

Thin epoxide resin or plastic strip-off backing film

Elongated reversal points for good bonding and low cross–sensitivity

Foil element

Alignment marks

Gradual change in cross-section eliminates stress concentrations

Large-area tags for soldering on leads

Figure 4.2 Typical foil-type strain gauge

Figure 4.3 Inductive sensors used for a door-mounting application (*Courtesy KUKA Welding Systems and Robots Ltd*)

Strain gauges

Strain gauges are widely used throughout industry for the measurement of strain – the small alteration of shape due to forces or torques acting upon an object. Figure 4.2 shows a typical strain gauge which could be used on a robot gripper. The strain gauge consists of a thin foil element which is carefully glued to the gripper, or the object upon which the forces act. The application of force causes a small change in the length of the gauge, which results in a change of electrical resistance that is detected by a Wheatstone bridge. The output from the bridge is amplified and the voltage signal is sent to the robot controller. As the voltage output is proportional to the force being applied to the gauge, the amount of gripping force can be continually measured until the desired force is reached. At this point the controller will stop the signal to close the gripper. By use of a strain gauge a single gripper can therefore be used to pick up various objects requiring different gripping forces.

Pressure-sensitive materials

These materials, which are usually supplied in sheet form, can be attached to the gripper or to the surface of the robot arm to detect contact with an object. When a force is applied to the plastic the resistance changes, the electrical output is fed to the controller which will either stop the robot or branch to another part of the program. These sensors are used primarily to indicate that contact or collision has been made between the robot and an object.

4.3 Non-contact sensors

Non-contact sensors gather information about the workpiece, the process or the environment without the use of physical contact. Some typical non-contact sensors used in robotic applications operate on the principles of electrical capacitance and inductance, sound waves (as in ultrasonics), light and laser beam techniques.

Capacitive sensors

The sensor head is equipped with an electrode system which generates a symmetrical electric field in the surrounding air. When a part is introduced into this field the dielectric constant of 1 is altered. The change in the dielectric value gives rise to a change in the current, which is measured using an electrode. Typical applications of these sensors are the locating of parts, part presence checks, classification of parts and their volume, and the tracking of contours.

Inductive sensors

When a supply voltage is connected to an inductive proximity sensor, an oscillator becomes excited and starts to oscillate. A concentrated and directed field is emitted from the oscillatory circuit coil. This is interrupted when a ferromagnetic body enters the field, causing a change in the current consumption. The difference in current consumption between the oscillating and the non-oscillating oscillator is used to trigger a defined switch point.

Figure 4.3 shows the end-effector for a door-mounting application. In this instance, three inductive proximity sensors are used to sense the translational offset in the X, Y and Z directions (by detecting reference surfaces on the car body next to the door aperture). The actual position is re-determined by the robot controller and the assembly program corrected accordingly.

Capacitive and inductive sensors are of straightforward and low-cost construction and can be used for tracking corrections in the case of simple workpieces. However, owing to their small working distances and their large volume, their use for robotic welding is highly restricted.

Ultrasonic sensors

Ultrasonic sensors work on the principle of either sending a single sound pulse, as in an echo-sounder, or emitting a continuous-wave signal where the phase displacement is measured between the transmission and the return of the signal. Either type can be used for the detection of objects and to measure distances. When using an ultrasonic sensor in conjunction with a robot, it is possible to set switching points so that when the workpiece reaches a defined distance from the robot a specified reaction is triggered.

Light-reflection sensors

These types of sensor operate by using white light or infra-red light directed onto the object which is to be monitored or detected. The light beam is reflected and detected by a receiver. The latter issues a signal which is converted into a processed signal, generally a DC voltage, that serves as an interrupt trigger to the robot controller. The robot, on receiving this signal, executes the relevant part of the program within the controller.

Laser distance sensor

The laser distance sensor allows the exact non-contact measurement of distances and the detection of component edges. The principle of the non-contact measurement of distances is based on the diffuse deflection of a laser beam off the surface of the workpiece, shown in Figure 4.4. The light collector is so designed that the resultant angle of reflection is proportional to the distance between the laser head and the workpiece. This is based on the principle of triangulation. The laser distance sensor is essentially characterized by the working distance (measured from the laser sensor to the work surface) and the measuring range (the depth, say, of the weld). The large measuring range enables substantial changes in structure to be

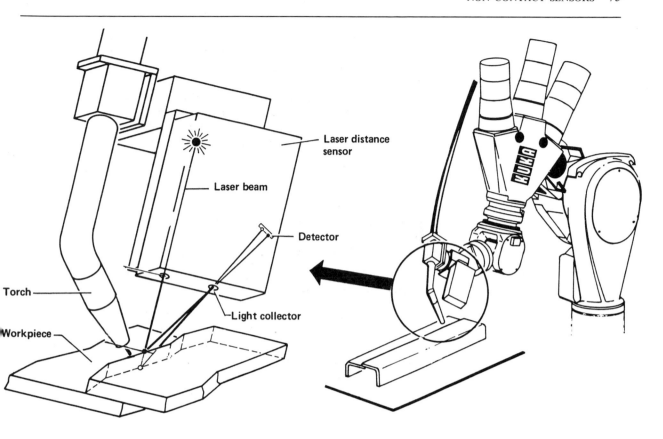

Figure 4.4 Laser distance sensor used for non-contacting measurement of distance, e.g. for locating workpiece edges in shielded arc welding, assembly, gluing etc. (*Courtesy KUKA Welding Systems and Robots Ltd*)

detected. Any changes of distance within the measuring range are sensed by a linear array of photodiodes on the detector.

Figure 4.5 shows a KUKA IR 160/15 robot with a dual-function tool for arc welding and stud welding. For arc welding, the robot uses a laser sensor to locate the beginning of the weld seam. On the basis of a command and actual value comparison, the controller calculates the offset required for the welding program concerned, which is then executed in the correct form. Figure 4.6 shows the same robot with the wrist twisted for stud welding. The studs are fed into the welding head automatically.

Laser scanning

Figure 4.7 illustrates the principle of laser scanning employed by KUKA. The sensor head is mounted on the advance side of the torch. A laser and a semiconductor line-detector sensor are integrated in the head for the purpose of detecting the seam. The measuring principle is the same as with the laser distance sensor. Two mirrors fitted to the same shaft are swivelled to and fro by a motor. The first mirror deflects the laser beam to the left and right of the seam, with the result that a strip of light is created across the joint. The second mirror reflects this strip of light onto the semiconductor sensor which stores distance measurements at the rate of 10 per second. The position of the individual picture elements is directly proportional to the

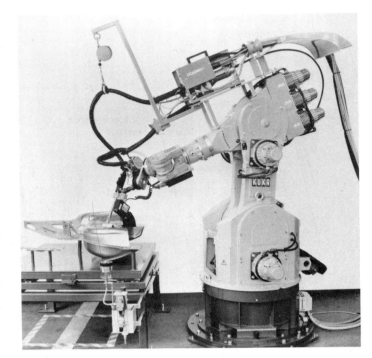

Figure 4.5 A KUKA IR 160/15 industrial robot with dual-function tool for arc and stud welding. Before the torch is used, the beginning of the weld seam is located with the aid of a sensor (*Courtesy KUKA Welding Systems and Robots Ltd*)

Figure 4.6 Stud welding with a KUKA 160/15 industrial robot. The studs are fed automatically. The robot switches the dual-function tool from arc to stud welding, depending on the program, by twisting its wrist (*Courtesy KUKA Welding Systems and Robots Ltd*)

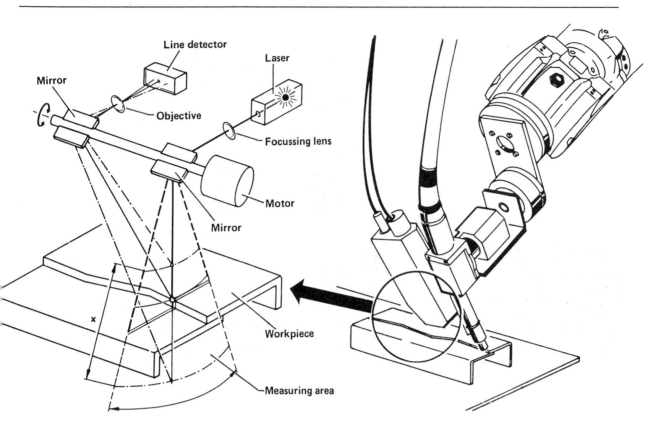

Figure 4.7 Laser scanner (*Courtesy KUKA Welding Systems and Robots Ltd*)

distance from the workpiece surface. It is therefore possible to determine not only this distance but also the start, centre and volume of the seam, by means of an image-analysing system. This provides on-line data for lateral, vertical and speed correction of the welding process to the robot controller.

4.4 Sensors for welding operations

Depending on their functional principle and design, sensors can perform all or some of the sensor tasks in automated shielded arc welding such as:

- Detection of the start of the seam
- Detection of the joint path and geometry
- Detection of the joint volume
- Detection of the end of seam

The environmental conditions encountered in shielded arc welding and the special requirements of the system that is to be automated give rise to a lot of restrictions. The three groups of sensors used for welding operations are shown in Figure 4.8:

1 Sensors based on mechanical probes (tactile sensors).
2 Sensors based on non-contacting measuring techniques.
3 Sensors based on the interpretation of welding process variables.

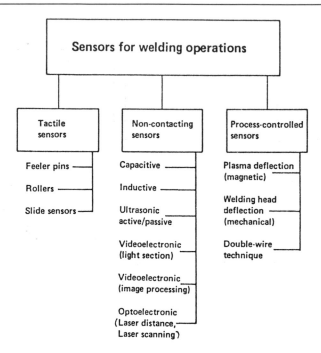

Figure 4.8 Sensors for welding operations (*Courtesy KUKA Welding Systems and Robots Ltd*)

Tactile (contact) sensors

These sensors are process-independent and inexpensive, but their use is restricted by their type of construction. In particular, they are liable to wear and can easily be damaged. Figure 4.9 shows a tactile two-dimensional sensor used on KUKA robots for the automatic fine programming of welded seams. The tactile two-dimensional sensor allows the automatic teaching of a roughly preprogrammed seam and automatic program adaptation in the case of minor seam variations. Deflection of the probe causes magnetic bodies to move past current-carrying magneto-resistors. The influence of the magnetic fields alters the internal resistances and thus the output voltages of the magneto-resistors. These output signals are processed in the robot controller for the purpose of correcting the path.

First, the auxiliary points on the path of the weld seam are roughly programmed on a master workpiece with the sensor fitted to the gas nozzle. These points are taught by the use of a teach pendant or by CAD coordinates. The tactile 2D sensor is then fitted to the welding torch and the program run with 'on-line path correction' sensor function activated. The sensor follows the actual path and sends the appropriate correction signals to the controller with the effect that the torch is guided along the centre of the weld seam. The new, corrected path points are constantly stored in the controller. It is then possible for the corrected program to be executed. Figure 4.10 shows a KUKA IR 160 robot using the arc sensor in the shielded arc welding of components.

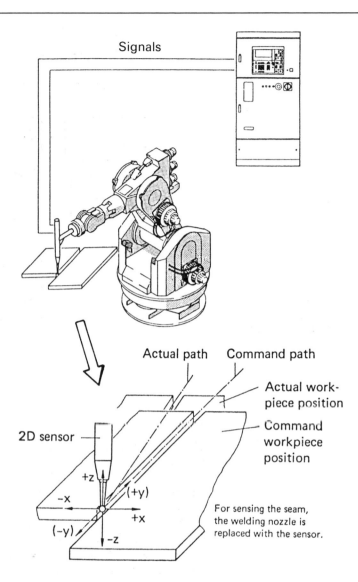

Figure 4.9 Kuka tactile 2D sensor for automatic fine programming of welded seams (*Courtesy KUKA Welding Systems and Robots Ltd*)

Process-controlled sensors

These sensors detect the path of the seam during the welding process. The arc sensor acquires data for tracking the weld seam by directly measuring the current, which depends on the length of the arc. This method affords the considerable advantages that the sensor does not have to be located on the advance side of the torch and that there is no signal delay. Moreover, no major problems are caused by dirt and heat since the arc itself functions as the transducer. This measuring technique is, however, restricted to vee and fillet welds and also to minimum current level and leg lengths.

Figure 4.10 A Kuka IR 160
robot using the arc sensor
(*Courtesy KUKA Welding
Systems and Robots Ltd*)

Non-contact sensing

For the non-contact sensing of a weld seam path in three dimensions, laser sensing is currently attracting the most interest for research and development. With their compact design, large measuring range and working distance and also small measuring area, laser sensors enable surface structures to be detected with the aid of the search function of the robot. Laser scanners are well suited to tracking the path of a weld seam as well as determining its start position and volume. They are particularly suitable for use in the welding of thin sheet metal and currently represent the most comprehensive sensor concept for shielded arc welding applications.

4.5 Tactile sensors for assembly and handling operations

Figure 4.11 presents an overview of sensors used for assembly and handling tasks.

Tactile sensors such as probes and feeler pins are used not only for locating workpiece edges and component positions but are also capable of functioning as collision-protection devices, e.g. the shutdown of a system when a force limit is reached.

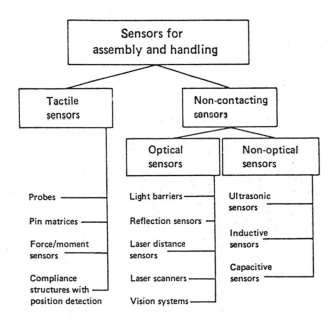

Figure 4.11 Sensors for assembly and handling operations (*Courtesy KUKA Welding Systems and Robots Ltd*)

Pin matrices which detect the distribution of forces on surfaces and in structures provide information on part characteristics and part positions. They have been employed only very rarely to date and are still in the prototype stage. The various tasks in question can often be performed better by a visual sensor.

A great deal of use is made of one- and two-dimensional force/moment sensors in present-day assembly operations. They are mostly employed for monitoring limits on assembly forces or for stopping the robot motion when the programmed force has been reached. Compliance structures are of great importance for assembly operations. The tolerances are compensated for by means of resilient mounting arrangements. These sensors enable the assembly procedure to be carried out quickly.

Force sensor

The force sensor measures the relative motion between two components with the aid of a measuring resistor. This relative motion is evaluated to determine the forces between the gripper and the workpiece, to detect the presence of the workpiece or to ascertain any malfunction.

The sensor is fitted between the robot arm and the gripper. When relative motion occurs between the sensor components, a small steel plate attached to the arm is moved relatively to the measuring resistor, which is attached to the gripper. The size of the workpiece is determined on the basis of the resistance measurement resulting from the relative motion of the sensor components.

Tactile 3-force/3-moment sensor

The sensor shown in Figure 4.12 measures both force and moments. It can be used wherever forces and moments acting on the end-of-arm tooling of the robot have to be detected and controlled. This is necessary, for instance, in a deburring operation, where the pressure is measured and adjusted, or in assembly work, where a snug fit is achieved with the application of a

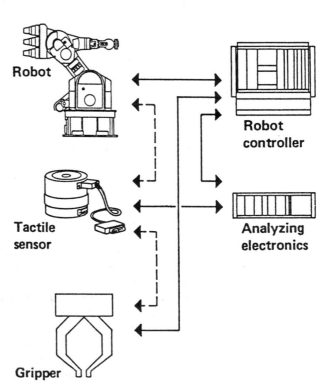

Figure 4.12 Tactile sensor with analysing electronics in an integral electric and mechanical unit (*Courtesy KUKA Welding Systems and Robots Ltd*)

defined force. The robot controller, aware of the forces and moments that are required, monitors the motion of the robot and ensures the reliable execution of the work even when the components are subject to variation.

The sensor is positioned between the robot wrist and the tool; all forces and moments acting on the tool are detected by means of strain gauges and converted to electrical voltages. A microcomputer in the sensor head calculates and digitizes these voltages and transmits them to the analysing electronics where they are resolved into three Cartesian coordinates. This data is then sent to the robot controller for program execution.

4.6 Robot vision systems

Perhaps the most complex sense that humans possess is that of vision. One glance at an object will gather such information as its shape in three dimensions, its orientation, colour and some estimation of its speed and position relative to the surroundings. Upon gathering this information the brain carries out a recognition check and calls on previous experience to decide on what action to take. Obviously robotic systems are unlikely ever to approach this degree of sophistication. However, even a simple vision system can enhance a robot's capability and enable it to carry out tasks that would otherwise be impossible for it to perform.

Since the mid-1970s there has been considerable development of vision systems for robotic applications, with an increasing number of highly sophisticated automated systems being successfully employed in industry. In particular, in the automobile industry vision systems are now being used for such applications as windscreen insertion, body gauging, wheel assembly, welding and seam sealant systems.

The great advantage of robot vision systems over the use of conventional sensors is that a vast amount of data can be collected and processed from just one instantaneous viewing. Hence, they offer performance at process speeds and so provide information to institute corrective action before the process goes out of control.

Binary-image processing

A binary-image processing vision system consists of four units: the camera, the analysing electronics, an operating unit and a monitor.

The cameras in use today are those that utilize semiconductors, because of their robust construction, small size and high optical quality. They have the additional advantages that there are no geometric errors in the image scanning, they operate at 50 frames per second and have a square arrangement of pixels in the operating zone. A pixel is a single element of a matrix sensor array, i.e. a photosensitive element. Pixels can be arranged in a single line, a square or a rectangular configuration, depending on the application. For instance, a single array could be 1×50 pixels, and likewise a two-dimensional array could be 150×150 pixels.

In the analysing electronics, the camera signal is reprocessed into a binary image (i.e. one expressed in ones and zeros) of the parts, which represents their significant features, e.g. shape, area, number and position of holes. This is achieved by sensing a high contrast object against its dark background, increasing the brightness to detect unevenly illuminated parts, or detecting the edge of an object which protrudes distinctly from the rest of an unclearly defined part.

The visual sensor is controlled by means of the operating unit. The sensor has to be calibrated to enable it to work in conjunction with the robot. The parts that have to be identified can then be taught into the analysing unit.

The monitor serves to display the TV picture and the binary image resulting from the preprocessing. The acquired data, e.g. the numbers of the identified parts, position of the area centres, rotational position, can be superimposed on these images and the parameters that have been set for the image preprocessing and feature recognition can be read off.

Applications of vision systems

Non-contact car body gauging of the Rover 800

Automatix have designed and installed a vision-based laser gauging system for measuring the critical dimensions of the complete body shell for the Rover 800. Figure 4.13 shows the vision body gauging cell at Austin Rover.

The system incorporates 62 laser and camera modules mounted within a structurally stable, vibration-free frame, through which the car body passes. Before the gauging process begins, the actual position of the car body on the jig is established by using three cameras which impinge on two sides of the car; these determine the coordinate information on the extremities of each panel. The data processed from these cameras is used to re-orientate the computer mathematical model. A total of 148 dimensions and body shell characteristics are taken, these include gauging front and rear screen apertures, engine and boot (trunk) compartment widths, sizes of door apertures and seal conditions, and the general geometric integrity of the side panels. These dimensions are compared to a database-generated mathematical model to an accuracy of 0.1 mm. A Statistical Process Control report is generated for every car. Dimensional details of car bodies are stored within the system file and this information is used to check for trends to ensure that cars are built to the correct tolerances.

The system uses an extensive suite of menu-driven modular software and is capable of determining that every vehicle is of the correct shape, that all apertures are the correct size and relative position to each other, that all door seal flanges and window apertures are correct and, finally, that all critical dimensions of the vehicle size are checked and compared against a mathematical CAD database-derived model stored within the Automatix computer system. The gauging system is capable of accepting model variations of the car and it completes the on-line inspection within 43 seconds. Until the advent of this system, it was not possible to check each car for dimensional stability and conformance to its computer-designed characteristics, because of the amount of time necessary when checking with conventional off-line contact gauging equipment: only limited inspection was possible for one or two car bodies per shift.

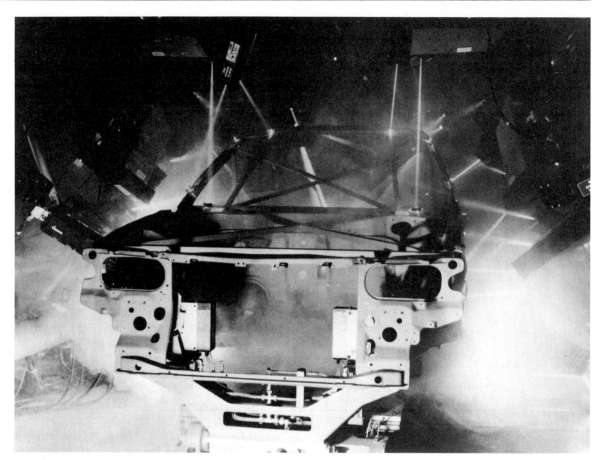

Figure 4.13 The vision body-gauging cell on the Austin Rover Group's assembly line (*Courtesy Automatix International UK Ltd*)

Direct car glazing systems

In 1983 Austin Rover unveiled its revolutionary fully automated glazing system for the Montego car line at Cowley, UK. The system was designed and installed by VS Technology, and is still operating today. In this application a Unimation Puma robot is used for the application of a chemical primer to the windscreens and for the application of the adhesive bond. This preparation station is situated above the windshield fitment area which is located by the moving track. The windscreens are passed down via a vertical conveyor to two Unimation 4000 robots which fit the front and rear glazing. Each robot gripper assembly consists of a glass retention system, i.e. four vacuum cups, and also holds the four vision cameras and the stepper motor drives used for glass adjustment and finally the insertion. The high payload on the robot wrist has caused some robot component wear; this has resulted in compounded inaccuracies in the vision analysis, which has caused some periodic failure in the system.

A 'second generation' fully automated glazing system for front and rear windshield insertion has been developed by Automatix for the Rover 800 line. The system employs vision-adaptive robotics, computer-linked charged coupled device (CCD) cameras with lasers and advanced mathematical computation techniques. Vehicles entering the glazing cell are checked to determine their precise position by two vision computers using cameras and lasers, which are mounted on a frame above the car bodies. At the same time two robots with vacuum grippers collect ready-beaded glass from the preparation system. When the front and rear aperture positions have been identified, the vision computers inform the two robot controllers. These are interfaced to two KUKA IR 662 robots which move to the windshield gauging positions, where glass size is determined and the style is checked (Figure 4.14). The vision computers calculate the perfect fit position for the glass and the robot controllers instruct the robots to carry out the installation movements. As the robots move back to await reinitiation of the cycle on the next car, the vision controller carries out a quality control analysis before the vehicle is released from the cell.

Figure 4.14 Automated glazing system on the Rover 800 line, incorporating Automatix vision systems and Kuka robots (*Courtesy KUKA Welding Systems and Robots Ltd*)

Figure 4.15 Automatic wheel-mounting with binary-image processing visual sensors (*Courtesy KUKA Welding Systems and Robots Ltd*)

The whole process takes 40 seconds, achieving considerable time savings over traditional manual methods and an unparalleled degree of accuracy. The system is capable of handling three variants of car body design and can execute the alignment and fitting of the glass to better than 0.5 mm accuracy.

Automatic wheel-mounting

An example of binary-image processing is the automated mounting of wheels in automobile production with robots and visual systems. The large positional tolerances of the wheel hubs, the undefined rotational position of the hole pattern and the continuous conveyor motion make high demands on the sensor technology employed. Figure 4.15 shows a wheel-mounting system which is installed in a production shop and is used to fit up to 5600 wheels a day, equivalent to an output of 1400 cars.

On each side of the conveyor, the system comprises:

- A wheel feed system with an integrated facility for the automatic insertion of bolts to the wheels.
- A robot equipped with a multiple nutrunner integrated in the wheel gripper.
- A sensor system featuring a video camera and a binary image-processing computer which determines the correction values for the robot motion program.

On both sides of the conveyor there is a tracking system for the car body with a built-in synchronizing transducer for coupling the robot with the continuous conveyor motion.

The hub patterns are binarized and filtered by the sensor and the position, area and contour length of all the image components are computed. With the aid of a geometric model of the hole pattern, all the image components which are not identical with the bolt holes are rejected. The location and rotation of the hubs are then calculated on the basis of the position of the bolt holes.

In this system, the front and rear wheels on both sides are mounted in a cycle time of 36 seconds. The torque and angle-of-turn are monitored during the tightening of the bolts. All 16 torquing operations on each car are documented by a printer connected to the system, which means that an immediate check on the quality of the bolt securing is incorporated into the production process.

Seam-sealing system

Another application of vision systems on the Austin Rover Group's Rover 800 Series production line is the seam-sealing system, designed and built by Automatix. Computer-linked stereoscopic 3D vision techniques enable adaptive automation to be applied in the seam-sealing cell for the application of mastic to underbody seams.

When a car bodyshell enters the cell it mechanically aligns to within ± 30 mm of its nominal position. A vision computer establishes the vehicle's precise position by using stereo vision algorithms calculated from four sets of stereo pairs of cameras focused on four master build-holes in the underbody. Six Unimate Puma 761 robots that apply the sealant are informed by the vision computer of the bodyshell's exact location and the master programs are offset to apply the mastic correctly. Any differences in the underbody styles, such as lefthand/righthand drive, are automatically identified and the robot program is adjusted to suit the particular model.

Using flow control and flow monitoring equipment, the robots apply a consistent 25 mm width and 2 mm depth of mastic accurately over the seams, completing the cycle within 40 seconds.

The benefits of this system include unrestricted access to the whole of the underframe and a reduction in material costs, since the application of the mastic is measured and monitored. Also, due to the accuracy of the system

(± 0.5 mm) and the nature of the body positioning, ideal application angles have been achieved which ensure seam coverage and enhance the seam penetration. This has resulted in improved quality of application, thus preventing water penetration of the seam and premature body corrosion.

Problems

1 Explain why sensors are used as part of the robot and peripheral devices.
2 Explain the operation of three types of non-contact sensors and where they may be used in a robot cell.
3 Sketch a foil-type strain gauge; explain the principle of operation and how it is temperature compensated.
4 Piezo-electric sensors are often used to determine if a part is present in the gripper; explain how this type of sensor operates and its advantage over other tactile sensors.
5 Explain, with the aid of sketches, the principle of using laser scanning to track the edge of a component.
6 What part do sensors play in the task of welding? Explain the groups into which they fall.
7 Explain how process-controlled sensors operate.
8 Sensors used for assembly operations differ from those used for some other operations; outline some of the special problems that are inherent in this application.
9 State the advantages that vision systems have over conventional sensors.
10 Explain the process of binary image processing and give some industrial examples of the use of this type of vision system.

5 Programming robots

The programming of robots can be a relatively simple or a highly complex operation. This depends upon the type of robot, the nature of the work it has to perform and its interaction with the environment and with peripheral devices. The task is compounded when more than one robot is working within the manufacturing area. A typical example is the automobile industry where robots are used for welding and spray painting car bodies. The car body shells arrive at random with a possible 12 variations of car body design or specification. Similarly complex programming is involved in the application of vision systems which are used for automatic windscreen insertion and car body inspection.

Before the programmer can start to plan the robot operation or write a flowchart he or she must be fully aware of the robot's limitations, the programming and logic functions that exist within the manufacturer's control software, and any software options that may be included within the controller.

Although all industrial robots vary with regard to their programming language and the inputting/outputting of information, most have the capability to perform the functions outlined below.

5.1 Robot software functions

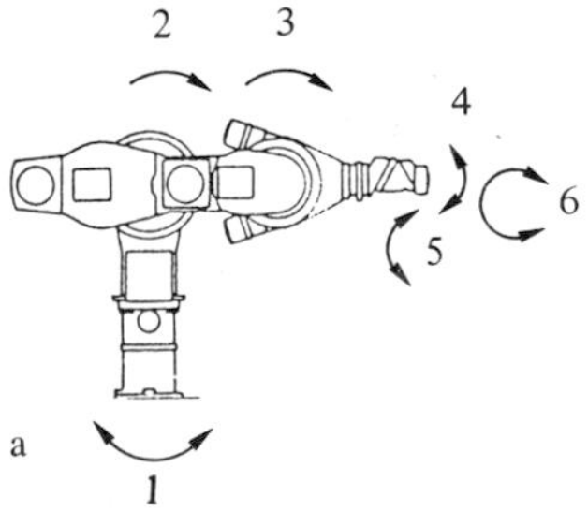

Figure 5.1 Coordinate systems for robot manipulators (*Courtesy KUKA Welding Systems and Robots Ltd*)
a Single-axis movement (6 axes)

Coordinate systems

Programming the moves of a robot is made simpler by the provision of various coordinate systems. These are shown in Figure 5.1.

Axis coordinate system

By means of the teach pendant pushbuttons, each individual axis can be moved in a positive or negative direction.

Cartesian robot coordinate system (world coordinates)

By means of the teach pendant keys a movement of the tool centre point in the basic coordinate axes X, Y and Z can be achieved, whereby the tool keeps its orientation in space constant.

Cartesian tool coordinate system

A Cartesian coordinate system is used which references the end-of-arm tool centre point, defined by the user, to a Cartesian tool coordinate system. The tool coordinate system can be translated and rotated with respect to the world coordinate X, Y, Z system.

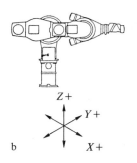

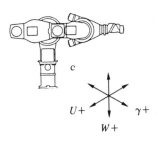

b Movement in Cartesian coordinates (world coordinates) where the centre is fixed at the robot datum point
c Tool coordinate system

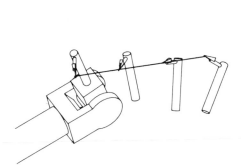

Figure 5.2 Linear position control

Position control

Point to point

The robot moves to the desired point specified within the program. With this type of move the robot takes the easiest path, i.e. with an articulated-arm type robot the movement of each joint follows the arc of a circle, so care must be taken to ensure that this unpredictable path does not impinge on any objects.

Linear position control

All axes move simultaneously to give a linear move to the desired point. In linear motion, the robot may stop momentarily on reaching the programmed point. Figure 5.2 illustrates linear position control. The linear path of the robot to a point can be adjusted by fine or coarse control when using high feedrates or when continuous motion is required (Figure 5.3).

Circular control

To program the end-effector to move in a curved path, three reference points are specified: the start point, the end point and some point between them on the circumference. The direction clockwise or counterclockwise is determined by the distance from the start position to the second position.

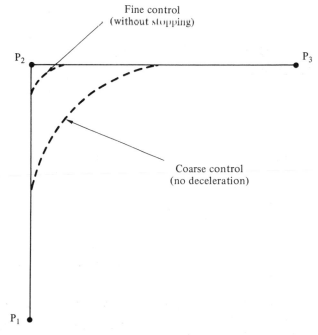

Figure 5.3 The path taken by a robot using either fine or coarse control around a point

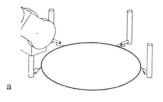

a

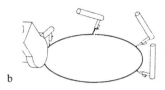

b

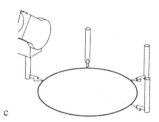

c

Figure 5.4 Hand/tool
orientation control
a Hand/tool orientation
correction relative to the start
and end positions
b Hand/tool orientation based
on all positions entered along
the circumference
c Movement along the circular
path with constant hand/tool
orientation to the circular path

Other control functions

Hand inclination and direction control

The wrist movement is controlled to maintain a specified angle and direction while keeping the tool tip at the same point; this is particularly important when using a robot for welding. Figure 5.4 illustrates hand/tool orientation control.

Tool centre point (TCP)

This control moves the coordinate system from the flange mounting on the end of the arm to the tip of the tool. Generally three tool geometry coordinates are entered into the robot control data, i.e. the tool tip X, Y and Z values which are measured from the mounting flange centre-line. Each tool offset can be called up within the program when the robot arrives at the automatic tool change station, or when the multi-tool end-effector is indexed.

User frame definition

Figure 5.5 shows how the robot Cartesian coordinates can be allocated to suit the orientation of different workstations; most robots can hold up to five different coordinate systems.

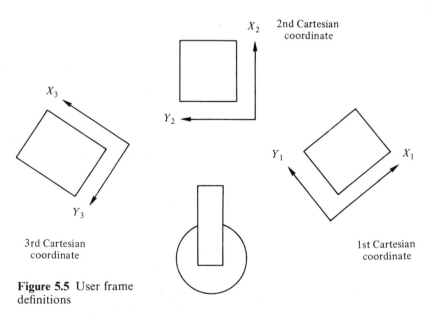

Figure 5.5 User frame definitions

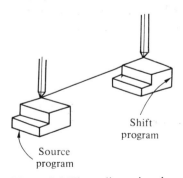

Shift program

Source program

Figure 5.6 Three-dimensional shift (parallel)

Three-dimensional shift

Programs to be executed on workpieces having the same configuration can be re-orientated by using the three-dimensional parallel and rotational shift, as shown in Figure 5.6.

Mirror image

This function enables a program to be rotated symmetrically around an axis plane. Also incorporated is the facility for program rotation with mirror imaging, as shown in Figure 5.7.

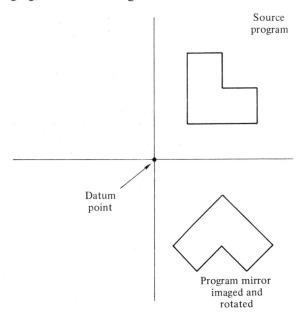

Figure 5.7 Program rotation with mirror imaging

Incremental input

This control function clears the position data of a robot point to zero, thus enabling the robot's next position to be taught as degrees of movement of each axis or, more easily, as a linear move in mm or metres.

Palletizing

By inserting a few points, i.e. the height of one column of components, the end position of the last stack, the number of stacks and the number of rows, a palletizing or depalletizing routine can be executed. This obviously reduces the amount of programming time and can be applied to components stacked in straight rows, in a 3D matrix or positioned in an arc pattern.

Reference point return

This control automatically returns the robot to a predetermined position at the end of a specified number of moves or at the program end.

Weave control

This function is used for generating tool tip patterns for arc welding and for adhesive and sealant depositing.

Program copy

Robot command data can be copied from one part to another part of the program, or the entire program can be copied to a different program.

Skip function

This allows the programmer to skip part of the robot program and go to another part of the program. This is particularly useful in conjunction with peripheral input signals, e.g. if a part is not present the robot will 'skip' to execute a different operation.

Subroutines

The robot controller can store a number of different programs, depending upon the memory capacity. Within a particular program, part of another program can be accessed, executed and on completion returned to its position in the original program. The relevant subprogram can be called up by using the output signals from the component's or workpiece's detection/identification sensors.

Registers

Registers are storage areas within the robot controller that can be used to execute mathematical functions, i.e. to add or subtract from a number stored in the register or to count the number of times an operation or cycle has been executed. The number of parts stacked on a pallet can be counted and also the number of repeat moves executed within a program.

Registers can also be used to control input and output data, e.g. when a robot has performed a task a certain number of times an output signal can initialize a peripheral device.

Setting data

Data held in the robot controller's memory in the form of binary or decimal numbers can set constants, flags or special functions. These might be the dimensions for setting up the tool centre point (TCP) for various end-effectors, or the accessing of signals from peripheral devices. Such data can also be used to call up different Cartesian frame references where four or five different workstations are being used. For example, pallets may hold various components that require different robot programs to be executed on recognition of a bar code or other part/component number identification device.

Binary settings

Binary settings enable a component to call up the required robot program within the controller. Input is in the form of binary numbers which may be signalled by a bar code positioned on the component.

Robot communications

Where a number of robots are used in a confined space, mathematical data relevant to the robot joint motor positions are exchanged between the controllers, so that when a collision is imminent one robot goes into 'hold'. This is sometimes referred to as 'cross talk'.

5.2 Planning the programming for a robot task

The robot program effectively controls all movements and sequences within the robot cell, from the physical manipulation of the arm to the processing of input/output signals to and from peripheral devices. It also monitors operations, e.g. welding current and wire feed in arc welding. The development of the program can be broken down into various stages, as shown in Figure 5.8.

End-effector requirements

The correct gripper or end-of-arm tooling must be selected to perform the

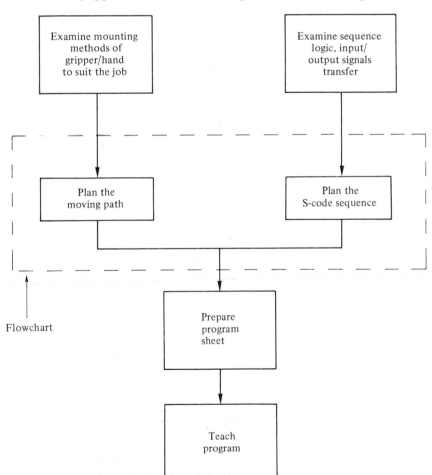

Figure 5.8 Program planning procedure

required task. Mounting arrangements for the gripper must be determined, the components or work orientation examined and methods of securing onto pallets or work tables decided. To ensure that the end-effector can reach the required positions, it may be easier to mount the component on the robot arm and take it to the operation rather than having the component static. If multiple tools are to be used, as in an assembly operation, it may be more convenient to have the tools indexing at the end of the robot arm, than to employ a separate tool-changing station.

Sequence logic

Sequence logic determines the sequence of the events that make up the program, in particular the order in which input and output signals are sent out and received by the peripheral equipment. For example, a controller sends an output signal to advance a conveyor to a predetermined position within the working envelope; the robot picks up a part from the conveyor on receiving a signal that the part is present; if, however, the part is not present the robot may go into a 'wait' function, return to its 'park' position, or start to execute a different program. The programmer determines such a sequence within the program by using special commands sometimes referred to as 'robot service requests' (RSRs).

Control of the cycle time of the program is important to ensure that production schedules can be met. The programmer knows the time taken for the conveyor to reach the desired position and must ensure that the arm is ready to grasp the part on arrival. Time must also be allowed for the gripper to open and close onto the component: these 'wait' commands are part of the robot's program.

Figure 5.9 Flowchart symbols

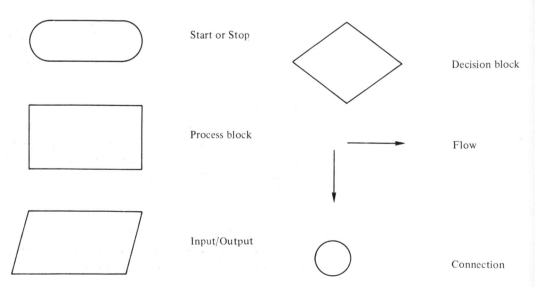

Planning the robot path

The robot path can be programmed in a number of different ways – point to point, continuous path, real world coordinates – according to the nature of the robot task. In all cases the programmer must determine and plan the quickest and most efficient route for the robot. A number of factors are considered:

- Possible collision with stationary or moving objects: it may be necessary to insert extra points in the program, some of which can be deleted when proving the program.
- The speed of movement between task elements: this may be able to be increased after a trial run, e.g. when lifting from the component to the next touch-on or approach position.
- The type of path that the robot has to follow: linear, circular interpolation, weave, coarse or fine position control onto or around a point; the position of the robot program insertion point relative to its datum; the feedrate at which the robot has to execute the path movement – although modern robots have high repeatability of the order of ± 0.2 mm, their accuracy decreases with increasing feedrate.
- The orientation of the robot end-effector, which must be correctly positioned for the approach to the component/work.

Planning the S-code (service code) sequence

Service codes give robot programs versatility since they are used for wait and gripper commands, manipulation of data in registers, branching to labels, turning on digital input and output signals, accessing welding data and palletizing commands etc. The programmer who is fully conversant with these codes can write a logical and well structured program.

5.3 Flowcharting for robot programs

In order to write a logical program the programmer should develop a flowchart before entering robot positions into the controller. The flowchart contains information on the robot path commands, S-codes and decisions that need to be made for successful program execution.

Flowchart symbols

The basic flowchart symbols can be found in any computer programming book. They are aids to planning the program so that all of the commands, input/output signals, decisions and instructions can be developed into a logical sequence of events. Figure 5.9 shows the symbols used and explains their meanings.

Example: a flowchart for a typical robot task

Two conveyors, A and B, transport the bases for two models of electric iron into a robot cell for application of adhesive prior to the assembly of the

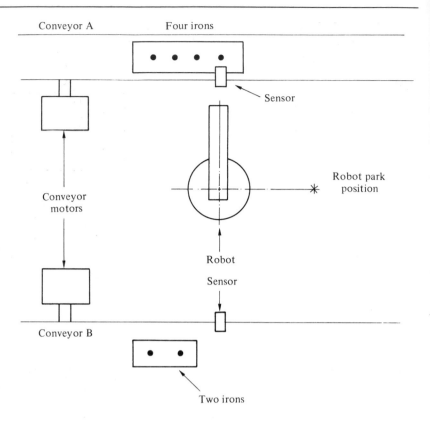

Figure 5.10 Layout for application of adhesive to electric irons

plastic bodies. They arrive in batches of four on conveyor A and in batches of two on conveyor B. If there are no bases present on conveyor A after an interval of 5 seconds, the robot will return to the 'park' position; if the irons are present the robot will apply adhesive to the first iron. The conveyor advances, the second iron has adhesive applied, and so on until the batch of four is completed. On completion, the robot will check if there are bases on conveyor B; if there are, adhesive is applied to the first electric iron, the conveyor is advanced and the second iron is processed. Failing the arrival of workpieces on conveyor B, the robot will return to conveyor A and the cycle will be repeated, provided that parts are present. The layout is shown in Figure 5.10.

In this example two input signals are required, one to indicate that parts are present on conveyor A and the other when parts are present on conveyor B. There are also two output signals, one for each conveyor, to advance the conveyor when a part has had adhesive applied. Two registers are used: one to count, and compare with the total of 4, the number of times that adhesive has been applied to irons on conveyor A, the other to do the same, with a total of 2, for the irons on conveyor B. The initial wait period of 5 seconds is entered into the program as an S-code.

Figure 5.11 shows a flowchart for robot moves and movement data for a robot performing this task. The flowchart begins with the 'start' symbol followed by the process blocks.

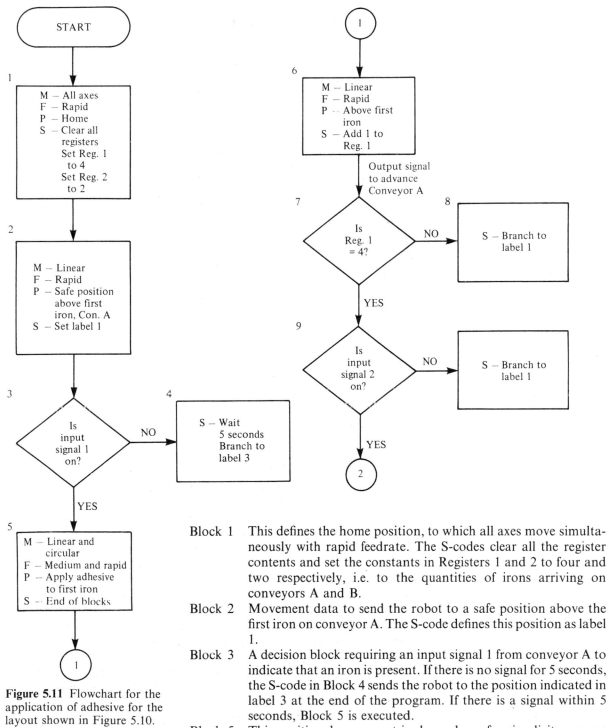

Figure 5.11 Flowchart for the application of adhesive for the layout shown in Figure 5.10. M = movement, F = feedrate, P = position, S = service code

Block 1   This defines the home position, to which all axes move simultaneously with rapid feedrate. The S-codes clear all the register contents and set the constants in Registers 1 and 2 to four and two respectively, i.e. to the quantities of irons arriving on conveyors A and B.

Block 2   Movement data to send the robot to a safe position above the first iron on conveyor A. The S-code defines this position as label 1.

Block 3   A decision block requiring an input signal 1 from conveyor A to indicate that an iron is present. If there is no signal for 5 seconds, the S-code in Block 4 sends the robot to the position indicated in label 3 at the end of the program. If there is a signal within 5 seconds, Block 5 is executed.

Block 5   This positional movement is shown here, for simplicity, as one block. In reality a number of blocks would be used, as this

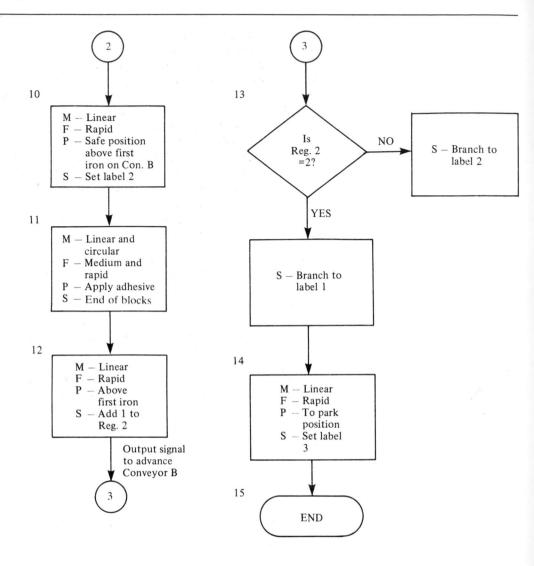

comprises all movements required to apply adhesive to an iron on conveyor A.

Block 6   This gives data to position the robot above the first iron; the S-code adds one to Register 1 and sends an output signal to the conveyor to advance.

Block 7   A decision block in which the contents of Register 1 are compared with 4: if the register shows less than 4, the S-code in Block 8 will return the robot to label 1 (i.e. the position in Block 2). If the register is equal to 4, the four irons have been processed and the program proceeds to Block 9.

Block 9   The robot is ready for a signal from input 2 to indicate that the parts are present. If the signal is not received the program branches to label 1, which is the position above the first iron on conveyor A, where it will wait for 5 seconds . . . etc. If signal 2 is received, Block 10 is executed; this is a move to a safe position above the first iron on conveyor B. Included in Block 10 is an S-code to set that position as label 2.

Block 11 Similar to Block 5: this positional information governs application of adhesive to the first iron on conveyor B.

Block 12 Data to position the robot above the first iron; the S-code adds 1 to Register 2 and sends an output signal to advance the conveyor.

Block 13 The register contents are compared with 2; if less than 2, the S-code branches the program back to label 2 for another execution of Block 10. When the register is equal to 2, the robot moves to label 1 which is the position contained in Block 2 above conveyor A, and the program repeats.

Block 14 This is a 'park' position near to conveyor A and is set as label 3.

Block 15 Program end.

This type of flow chart can be interpreted into any type of robot language since it represents the logical approach to a particular problem. Obviously some flowcharts can be very lengthy and may contain subprograms and branching to other programs within the robot memory. An example of the latter is a system employed by Autotech Robotics who design and manufacture Robotized Automation Systems. Steel pallets are used which have a matrix of accurately machined holes to accept various welding jigs which can be changed from pallet to pallet. The pallets are located on cones and attached to a conveyor which transports the jig into the robot working area. The job to be done by the robot is identified by a simple six-hole aperture code (similar to a bar code) on the side of the pallet. When the pallet, complete with the work ready for welding, is presented to the robot this code is read by a series of proximity sensors. The robot is thus informed of the program to be selected.

5.4 Teaching the robot program

Industrial robots can be taught movement and position data in three ways: lead by nose, teach pendant and off-line programming using computer techniques.

Lead by nose

This is the only method used to program paint-spray type robots. The experienced sprayer holds the spray gun attached to the end of the robot arm flange, and guides it through the series of movements required to carry out the operation. The taught positions are stored in the robot memory and are played back to spray the parts in production, exactly duplicating the sprayer's hand movements, technique, and dexterity. Figure 5.12 shows two Trallfa robots underspraying car bodyshells. This method of programming is sometimes known as walk-through or continuous-path control, as the position of the end-effector is recorded into the robot's memory at small increments of time (Figure 5.13). The programmed points can be scaled up if a model is used. This is advantageous when

Figure 5.12 Two Trallfa robots spraying the undersides of car bodies (*Courtesy ABB Robotics Ltd*)

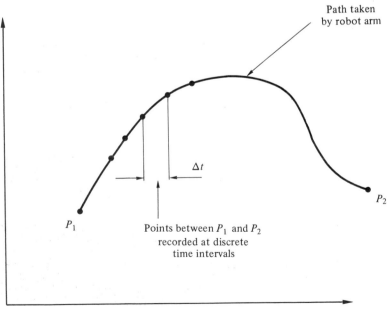

Path taken by robot arm

Δt

P_1

Points between P_1 and P_2 recorded at discrete time intervals

P_2

Figure 5.13 Continuous path control

Figure 5.14 Programming a KUKA robot for a drilling operation using the teach pendant (*Courtesy KUKA Welding Systems and Robots Ltd*)

programming a large robot which is difficult for the operator to manipulate physically.

Difficulty in programming this type of robot to achieve high accuracy and to follow straight-line paths has led to the advancement of electric drives to replace hydraulically-driven systems.

Teach pendant

Programming via the teach pendant is the most common method of entering data into the user memory of the robot controller. Figure 5.14 shows the programmer using a teach pendant for inserting a program into a KUKA robot to enable the robot to execute a drilling operation. Some industrial robots incorporate a joystick on the programming unit to

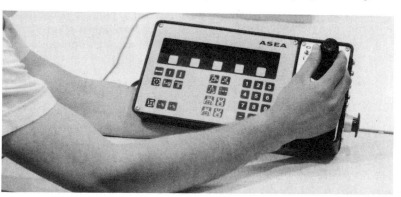

Figure 5.15 Operator using the joystick control to move the axis of an ASEA robot (*Courtesy ABB Robotics Ltd*)

control the robot movements. In Figure 5.15 the operator can be seen manipulating the arm of an ASEA robot using a joystick. This method of moving the manipulator reduces the operator learning period since three axes can be moved simultaneously. A teach pendant for the 600 FANUC robot is shown diagrammatically in Figure 5.16 to enable the layout of the pushbutton keys to be clearly seen.

The pushbuttons are subdivided into keys for jogging each axis of the manipulator and programming function keys that are used for inserting the type of move required, executing the program forward or backward, and single block execution. Some keys control special functions such as increasing or decreasing weld wire feedrate and voltage or handling operations such as opening or closing the gripper, or turning on and off a sealant gun. Visual displays can include a liquid crystal display for showing the block number, robot feedrate, welding voltage and wire feedrate. Indicator lamps are used to show the mode of operation, the coordinate system being used, and various hold, wait and alarm signals.

Some of the above functions are explained in greater detail in the following sections.

Modes of operation

There are generally three modes of operation: the teach mode, the repeat mode and the remote mode, the relevant mode being indicated by a light on the teach pendant. The teach mode allows the operator to enter positional data by jogging the manipulator axis to the required points; the axis feedrate can also be inserted, although this is generally 25% of the actual feedrate programmed when in the teach mode.

The repeat mode enables the programmer to repeat the taught moves continuously either forward or backward through the program. The programmed feedrate of the axes can also be overriden by means of the feedrate plus or minus key. In this mode the robot can be moved in a single block sequence either forward through the program or backwards. This is particularly useful since it allows the programmer to examine closely the moves to and from the component. In this mode all input/output signals, wait commands and other service codes are inoperative.

The remote operation is only valid when the teach pendant is switched off and the operator's panel is enabled. The program is executed at the programmed feedrate and all peripheral signals and service commands are executed; the robot will stop only on receipt of a wait command, hold or emergency stop signal.

Coordinate system selection

The light signals at top right of the pendant indicate the type of coordinate system that is being used for programming, i.e. jogging the robot by one axis at a time, moving in the Cartesian coordinate system or using the hand coordinate system. The relevant indicator lamp will remain illuminated while the coordinate system is selected. The 'mode' key allows the

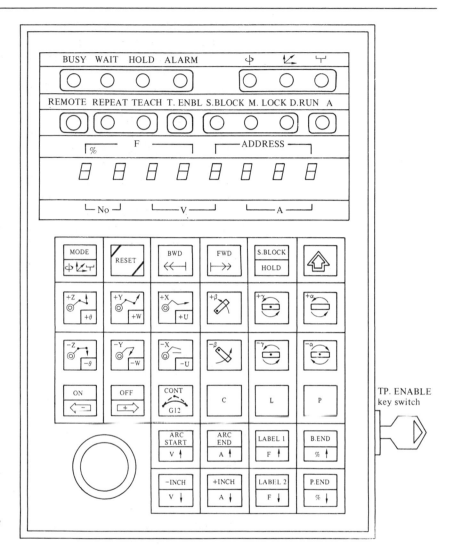

Figure 5.16 Fanuc robot teach pendant for the RC Controller (*Courtesy GMFANUC Robotics (UK) Ltd*)

programmer to change from one system to another; as a safety measure, the feedrate automatically defaults to zero when changing from one coordinate system to another.

Jogging keys

On the teach pendant there are 12 keys for moving a 6-axis manipulator (10 for a 5-axis manipulator) to the maximum or minimum position of each axis. For example, +Z moves the Z axis up, −Z moves it down; +X and −X move the *X* axis to or from the robot centre and +Y and −Y move to the left or right. The six other keys relate to movement of the wrist, giving plus or minus wrist swing or rotation.

As a safety precaution, the robot controller parameters can be set so that both the shift key and the relevant jog key have to be depressed to cause the arm to move. When a program has been entered as single-axis moves and it is played back in Cartesian coordinates, the axis limits may be overextended and cause an 'overtravel alarm' condition: this signal, which causes the robot to stop, is generated by a micro-switch which is actuated by a mechanical stop on the manipulator axis. The programmer has to press the reset key to reset the alarm, and jog the relevant axis away from the microswitch. The reset key on the teach pendant can also be used for resetting alarm signals from interlocking gates, light curtains or any peripheral device that sends the robot into a 'hold' or 'emergency stop' condition.

Programming keys

There are a number of programming keys on the teach pendant to enable the programmer to enter the desired path of the manipulator: P – move all axes to a certain point; L – move to a point in linear control; C – circular interpolation through a series of points. There are also block and end-of-program keys, and keys for entering the feedrate, which is indicated in the liquid crystal display. These keys are also used to increase or decrease the feedrate when the robot program is being executed in the repeat mode.

Off-line programming

Off-line programming techniques generally use 32-bit or mainframe computers to simulate graphically robot movements around a cell or perhaps along a manufacturing line. The software packages are very sophisticated: the simulation can be shown either in a three-dimensional wire frame form or as a solid model with full solid shading. Once the robot program is created and verified by the simulation, the computer data is postprocessed into the required robot language and loaded into the robot memory for implementation.

One of the most successful software packages used in industry for off-line robot programming is GRASP (BYG Systems) which uses a combination of textual and graphical methods. The robot program containing all the logical information and input/output signals data is developed in text form. The positional steps are then entered using interactive graphics. The use of instant graphical feedback helps the programmer to get the positions absolutely correct. Collision detection is carried out either visually or automatically by the system and can be checked for clashes over a specified time period of the predefined robot program.

Some computer plot illustrations of GRASP are shown in Figure 5.17, where a Puma robot is being used for the loading of an automated guided vehicle (AGV) from a conveyor system, and Figure 5.18, which shows the computer visual display of a machining cell in which the components to be machined are picked up by a robot and loaded into the CNC lathes. Figure 5.19 shows a Reis robot welding intersections of pipes located on two workstations; this application of GRASP highlights the advantage of using this system of off-line programming, since the control allows each of the

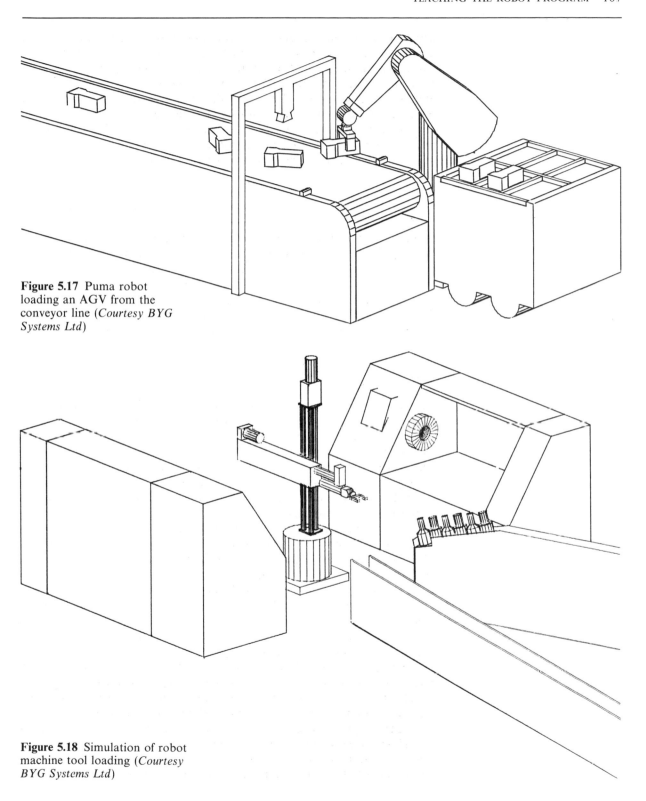

Figure 5.17 Puma robot loading an AGV from the conveyor line (*Courtesy BYG Systems Ltd*)

Figure 5.18 Simulation of robot machine tool loading (*Courtesy BYG Systems Ltd*)

jointed structures of the manipulator to be moved around accurately to produce the welding program.

Another off-line programming system called WRAPS is being developed by Kim Goh of Loughborough University for the generation of robot arc welding programs associated with an expert system. The system is used to generate the robot program complete with the best available procedure, selected from a database, which is then communicated directly to the welding system. WRAPS is also capable of receiving data from the welding system which will enable it to modify and eventually optimize the welding procedure on the basis of manufacturing experience.

One of the reasons for this development has been that, using the teach pendant, the application of robotics to the arc welding of small fabrication batches is limited: the introduction of new assemblies into production incurs the cost of delay while the robot is unproductive, and also the visual location of the welding wire tip can introduce significant positional errors.

Advantages of off-line programming

1 Improved robot and programmer safety:
   (a) The programmer is remote from a potentially hazardous environment.
   (b) The programmer and the robot are not placed at risk by the accidental operation of the wrong controls, as when the robot is very close to the workpiece, the jigs or the operator.
   (c) In off-line programming the operator's eyes need not be close to the tool point to achieve the desired accuracy of positioning. Conventional programming places the operator in a vulnerable position inside the working envelope. High-precision tasks require this risk period to be long and continuous, particularly if the program consists of several hundred points. Control mistakes, operator fatigue and errors induced by visual problems such as restricted view and eye strain can easily cause mistakes in the program and poor accuracy.
2 Postprocessors enable a variety of types of robots to be programmed from one workstation.
3 The system permits verification, assisted by the graphical simulation system, of robot programs in terms of positional data and program logic. This may include input/output signals for the control of peripheral equipment, the processing of sensory inputs, path control information and programming structures such as loops, branches and wait instructions, and the inclusion of previously written subroutines.
4 In off-line programming the robot program is developed, partially or completely, without requiring direct use of the robot, which can remain in production while its next task is being programmed. This makes the use of robots economic for small batch production, and increases the economic viability of the robot installation.
5 The use of GRASP enables complete automation systems to be planned, built up and modified to suit the desired requirements. This can include

the layout of machine tools, materials handling devices such as bowl feeders and conveyor systems, as well as the evaluation of different robots and their individual positioning and work envelope considerations.

6 Off-line programming allows data built up in CAD and production control systems to be incorporated into the design of the robot installation.

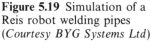

Figure 5.19 Simulation of a Reis robot welding pipes (*Courtesy BYG Systems Ltd*)

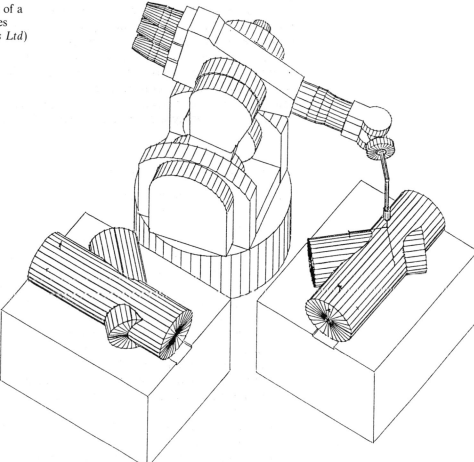

5.5 Executing a robot program and changing the memory data

There are two further functional devices whose operation the programmer must understand before embarking upon robot programming. These are the operator's panel and the CRT/MDI panel.

Operator's panel

Figure 5.20 shows the layout of the operator's panel for a 600 FANUC robot. The panel is an integral part of the robot controller: it enables the operator to have some control over the robot functions but not direct access to the program if the 'memory protect' is functional. The operator can, for example, start up the robot at the beginning of each shift, execute a

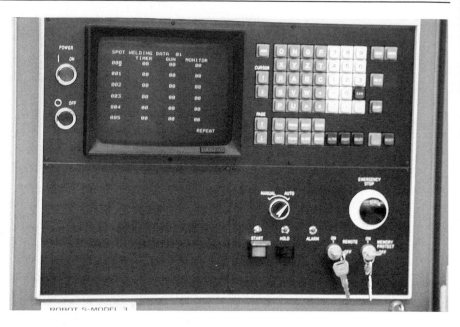

Figure 5.20 Fanuc RC control and operator's panel (*Courtesy GMFANUC Robotics (UK) Ltd*)

'hold' function if required and operate an emergency stop. The panel shown has the following functions.

Start button

This is pressed on initial start-up to execute the zero point return; once the relevant program has been called from memory, the operator can use it to start the operating sequence.

Hold button

This allows the operator to hold the movement of any axis during any stage of the program execution – say, to relocate a jig or to make adjustments to peripheral devices. The start button is pressed to restart the program.

Alarm lamp

This is illuminated when the robot is in any alarm condition, e.g. hold, emergency stop or when the safety barriers are broached.

Disable/Enable

To initiate zero point return and the execution of the robot program in single block and repeat mode, this has to be set to 'enable' mode. When set to 'disable', none of the robot axes will move nor will any service code operate.

Remote ON/OFF

With the remote control ON, only such functions as display, emergency stop, reset and single block moves can operate; all other functions are controlled by the teach pendant keys.

Memory protect

The purpose of this switch is to protect the program in memory from being erased or tampered with; usually only the programmer has access to the key.

Emergency stop button

This immediately cuts off all power to the manipulator motors, so stopping all robot movement. If this switch is connected to peripheral equipment this will also be halted until the button is reset.

Cathode ray tube/manual data input (CRT/MDI) panel

The MDI panel enables the programmer to input and edit data such as G-codes, feedrates, and service codes to the robot program. It is also possible to change setting data, registers and parameters by using the alphanumeric keyboard. Figure 5.21 shows the CRT/MDI panel for the 600 FANUC RC controller. The CRT displays positional data of the axes, the program contents (three blocks per page), registers, parameters, setting data, diagnostic contents, welding conditions and welding parameters, error codes and alarm conditions, e.g. emergency stop (dead man's handle). In addition, software menus are displayed for setting up frame references, scaling functions, mirror-imaging of moves, shift commands and setting welding data. The contents of the data memory can also be displayed. For example, the number of robot programs stored within the memory and their relevant program numbers and, in terms of program points, the total used area and the remaining vacant area.

The CRT/MDI panel is portable to allow for remote operation away from the controller or for connection to the controller of another robot cell for editing a program or changing various data.

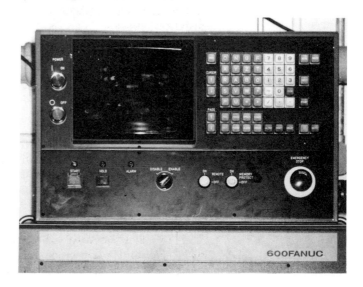

Figure 5.21 The CRT/MDI panel for the 600 FANUC RC controller (*Courtesy Bournemouth and Poole College of Further Education*)

5.6 Robot languages

Throughout the evolution of robotics, many different languages have been written to control the movement of robot manipulators or robotic-type actuators. As with most microcomputer-controlled equipment, there is no universal language, nor does one seem likely to be developed – one has only to look at CNC machine tool languages from APT onward, CAD/CAM commands, or programmable logic controller system data. Robot control languages tend to be equally diverse.

In the following simple example, the robot program for executing the task is given in three different languages used by well-known robot manufacturers.

Example: Feeding castings to an automatic bandsaw

A robot is used for feeding castings from a conveyor to an automatic bandsaw which cuts off the runners and the risers. On completion the robot removes the casting, places it on the outgoing conveyor and returns to repeat the operation. There is a 2-second wait when gripping and releasing the castings and the time taken to carry out the sawing operation is 6 seconds. The complete sequence is repeated 50 times.

Robot program for 600 *FANUC RC control*

In this format, linear moves are described with a G-code, i.e. G01 followed by a feedrate value from 0 to 2000 mm/s; point-to-point moves, G0, are displayed by a single digit feedrate, i.e. 0 to 8. Service codes commands (S) perform the logical sequence of the program: 'wait' commands, gripper instructions and the setting up of register data. Positional data blocks enable S-codes of only up to 4 characters to be entered; the block containing G98 is a non-motion block which makes 16 additional S-codes available. For clarity, the six-axis positional data is not shown for the above example, but a description is given of the end-effector recorded positions.

The program language below is followed by an explanation of the program blocks.

```
PROG 00001
N0000 G98 S01,1 S63,8,1 S63.9,3 S71,1 S00
N0001 F8 S00
N0002 F8 S97,1 S00
N0003 G01 F500 S00
N0004 G98 S60,20 S70,1 S00
N0005 G01 F500 S00
N0006 F8 S00
N0007 G01 F500 S00
N0008 G98 S60,20 S71,1 S00
N0009 F8 S60,60 S00
N0010 F8 S00
N0011 G01 F500 S00
```

```
N0012 G98 S60,20 S70,1 S00
N0013 G01 F500 S00
N0014 F8 S00
N0015 G01 F500 S00
N0016 G98 S60,20 S71,1 S00
N0017 G01 F500 S10,1 S00
N0018 G98 S17,1,50 S30,1 S32,2 S00
N0019 F8 S97,2 S99
```

The program number is defined as 1.

N0000   This clears register number 1. User frame number one is used, this is defined in setting data 8. Tool number 3 is stated in setting data number 9, the gripper opened and end of block.

N0001   Move to the park position.

N0002   Move close to the casting on the conveyor and define the position as label 1.

N0003   Move to pick up casting.

N0004   Wait for 2 seconds and close gripper.

N0005   Lift casting from conveyor.

N0006   Move near to the bandsaw.

N0007   Position casting in the bandsaw holding device.

N0008   Wait for 2 seconds and open gripper.

N0009   Move away from the bandsaw and wait for 6 seconds.

N0010   Move to the casting held in the bandsaw.

N0011   Move to grasp the casting.

N0012   Wait for 2 seconds and close the gripper.

N0013   Remove the casting from the bandsaw.

N0014   Rapid move to above the outgoing conveyor.

N0015   Position the casting onto the outgoing conveyor.

N0016   Wait for 2 seconds and open the gripper.

N0017   Move away from the conveyor in a linear path and add one to the data in register one, block end.

N0018   Compare the data in register one to the constant of 50, if it is equal to 50 branch to the position defined as label 2, if it is not equal to 50 branch to the position defined as label 1.

N0019   Move to the park position and define the position as label 2, program end.

Robot program for a Unimation Puma (VAL II) control

When programming the Puma robot the end-effector is taken to the relevant position and the six-axis positional data is recorded and defined as locations or names. The program is then written inserting these names as positions. The EDIT command is used to call the editor program and to give a name to the program, in this case BANDSAW FEEDING. The next instruction is to EXECUTE the program 50 times in succession.

```
.  HERE CASTING

X/JT1    Y/JT2    Z/JT3    O/JT4    A/JT5    T/JT6
54.35    236.78   -29.03   -202.6   78.90    113.76

.  HERE BANDSAW

X/JT1    Y/JT2    Z/JT3    O/JT4    A/JT5    T/JT6
305.5    45.78    35.8     -56.8    -34.98   78.21

.  HERE CONVEYOR

X/JT1    Y/JT2    Z/JT3    O/JT5    A/JT6    T/JT6
205.45   -389.4   25.97    -178.4   78.63    170.3


.     EDIT BANDSAW FEEDING .1
.     EXECUTE BANDSAW FEEDING .1,50
1.?   SETI COUNT = 0
2.?   10 APPRO CASTING, 25
3.?   MOVE 'S' CASTING
4.?   WAIT 2
5.?   CLOSE I
6.?   DEPART 'S', 50
7.?   APPRO BANDSAW, 100
8.?   MOVE 'S', BANDSAW
9.?   WAIT 2
10.?  OPEN I
11.?  DEPART 'S', 350
12.?  WAIT 6
13.?  APPRO BANDSAW, 100
14.?  MOVE 'S' BANDSAW
15.?  WAIT 2
16.?  CLOSE I
17.?  DEPART 'S', 250
18.?  APPRO CONVEYOR, 25
19.?  MOVE 'S', CONVEYOR
20.?  WAIT 2
21.?  OPEN I
22.?  DEPART 'S' 100
23.?  SETI COUNT = COUNT +1
24.?  TYPEI COUNT
25.?  GOTO 10
26.?
```

1 This tells VAL to set the value of the integer variable named COUNT to zero.
2 Approach to within 25 mm of the casting.

3 Move to the casting in a straight line, this is indicated by S at the end of the command.
4 Wait for 2 seconds.
5 Close the gripper onto the casting immediately.
6 Move 50 mm from the conveyor in a straight line.
7 Move to within 100 mm of the bandsaw.
8 Move to the bandsaw in a straight line.
9 Wait 2 seconds.
10 Open gripper immediately.
11 Move along a straight line for 350 mm away from the bandsaw.
12 Wait 6 seconds.
13 Move to within 100 mm of the bandsaw.
14 Move to the bandsaw to grasp casting.
15 Wait 2 seconds.
16 Close the gripper immediately onto the casting.
17 Move 250 mm from the bandsaw in a straight line with the casting.
18 Move to within 25 mm of the conveyor.
19 Move to the conveyor in a straight line.
20 Wait 2 seconds.
21 Open the gripper immediately.
22 Depart from the conveyor in a straight line for 100 mm.
23 This instruction causes the value stored in the variable COUNT to be increased by one each time the step is executed.
24 This displays the value of COUNT to be displayed on the terminal.
25 This restarts the execution of the program from 10, i.e. approach casting.
26 The carriage return indicates the end of the program.

Robot program for an ASEA robot

The program below is followed by an explanation of the commands.

```
PROG 1
 10    VELOC 500 mm/sec MAX VELOC 1000 mm/sec
 20    TCP 2
 30    RECT COORD
 40    POS V = 100%
 50    SET R5 = 0
 60    POS V = 100%
 70    POS 50% FINE 1
 80    WAIT TIME 2s
 90    GRIPPER CLOSE
100    POS V = 50%
110    POS V = 100%
120    POS V = 50% FINE
130    WAIT TIME 2s
140    GRIPPER OPEN
150    POS V = 100%
```

```
160    WAIT TIME 6s
170    POS V = 50% FINE
180    WAIT TIME 2s
190    GRIPPER CLOSE
200    POS 50%
210    POS 100%
220    POS 50% FINE
230    WAIT TIME 2s
240    GRIPPER OPEN
250    POS 50%
260    POS 100%
270    ADD 1 TO R5
280    JUMP TO 60 IF R5<50
290    END
```

The program is given a number, in this case 1.

10 Define the basic and maximum velocities in mm/s.
20 Tool Centre Point for tool number 2.
30 Move the robot in rectangular coordinates.
40 Move to the 'park' position at maximum basic speed.
50 Set register number 5 to zero.
60 Move towards the casting on the conveyor at maximum speed.
70 Move onto the casting using fine point control at 50% of the basic speed.
80 Wait for 2 seconds.
90 Close the gripper.
100 Move the casting from the conveyor at half maximum speed.
110 Move towards the bandsaw.
120 Places the casting into the holding device on the bandsaw with fine position control.
130 Wait for 2 seconds.
140 Open gripper.
150 Move away from bandsaw with maximum speed.
160 Wait for 6 seconds while sawing is in operation.
170 Move, with fine control, to pick up the casting from the bandsaw.
180 Wait for 2 seconds.
190 Close gripper onto casting.
200 Move away from bandsaw with casting.
210 Rapid move towards output conveyor.
220 Position casting onto the output conveyor.
230 Wait for 2 seconds.
240 Open gripper.
250 Move from conveyor with fine control.
260 Rapid move to approach casting on input conveyor.
270 Add 1 to register number 5.
280 If the register contents is less than 50, jump to line 60.
290 End of program.

Programming using the KAREL language

KAREL is a high-level language developed by GMF Robotics Corporation and can be used to program different robot models in a variety of applications, since it is based on the concept of generic programming and control. This means that a program written for a particular operation can also control different variations of that operation. For example, a KAREL program written for spot welding car bodies will adapt not only to changes in body style but also to different body positions on the assembly line. The robots used can be of any mechanical configuration – cylindrical, rectangular or articulated – as the positional data in the KAREL system is based on Cartesian coordinates. As explained earlier, these positions represent a location in space (X, Y and Z coordinates) and also the orientation (pitch, yaw and roll). These coordinates are generally measured from the user system of the robot (this is sometimes referred to as the 'user-frame'), which can be defined by the programmer for a particular application. The tool coordinate system is defined with respect to the robot arm end-effector mounting flange.

The KAREL programming language

The KAREL language is composed of the following groups of statements:

- Data-type declarations
- Data computation
- Control structures
- Robot motion
- Inputs/outputs
- Routines
- Interrupts and condition handlers

These statements are used to create a program which will control the robot motion, handle process control data and interface with sensors and communication devices.

Example of programming in KAREL

The task of the robot is to load and unload components into the chuck of a machine tool. The robot moves from the 'park' position to the opened chuck and grasps the part. The component is moved, via the 'park' position, to be placed onto the output conveyor. The robot then moves to the input conveyor where it picks up the next part for machining. Upon picking up this part, the robot moves through the 'park' position to insert the component into the chuck. The chuck is closed and the robot returns to the 'park' position. A count is made upon the completion of each sequence.

The program below is followed by an explanation of the statements.

```
 1 PROGRAM machine tool loading
 2
 3 VAR
 4 PARK            : Position     : park position
 5 CHUCK           : Position     : chuck position
 6 CONVEYOR IN     : Position     : conveyor-in position
 7 CONVEYOR OUT    : Position     : conveyor-out position
 8 PART COUNT 1    : Integer      : counter variable
 9
10 CONST
11 CHUCK OPEN  = 1    : output signal to open chuck
12 CHUCK CLOSE = 2    : output signal to close chuck
13 STOP CYCLE  = 3    : input signal to stop program
14
15 BEGIN
16 $SPEED = 500
17 $MOTYPE = linear
18 $TERM TYPE = coarse
19 PART COUNT 1 = 0
20 REPEAT
21       MOVE TO PARK
22       MOVE TO CHUCK
23       CLOSE HAND 1
24       DOUT [1] = ON
25       DELAY 2000
26       MOVE TO PARK
27       MOVE TO CONVEYOR-OUT
28       OPEN HAND 1
29       DELAY 2000
30       MOVE TO CONVEYOR-IN
31       CLOSE HAND 1
32       DELAY 2000
33       MOVE TO PARK
34       MOVE TO CHUCK
35       DOUT [2] = ON
36       DELAY 2000
37       OPEN HAND 1
38       DELAY 2000
39       MOVE TO PARK
40       PART COUNT 1 = PART COUNT 1+1
41
42       UNTIL DIN [stop cycle] = ON
43
44       WRITE ('Total Components Machined', part count)
45
46       END machine tool loading
```

The program commences with the program name and 'variable statements' which in the above program define the taught points. These are given variable names such as 'PARK position', 'CHUCK', 'CONVEYOR IN' and 'CONVEYOR OUT'.

Lines 10 to 13 are 'CONSTant' statements which in this case declare output signals on 1 and 2 ports from the robot controller for opening and closing the chuck. Also an input signal on port 3, from a peripheral device, will stop the robot cycle.

Line 15 indicates the start of the operational section of the program.

Lines 16 to 19 describe the actions to be performed by the robot, i.e. SPEED 500 mm/s MOving in a linear TYPE motion. Approach each position with coarse control and TERMinate motion at that point before executing the next statement and set PART COUNTer 1 to zero.

Line 20 is the beginning of the main program loop which ends at line 40, this will be repeated until digital input 3 is on.

Line 21 MOVEs the robot TO the PARK position.

Line 22 MOVEs the robot TO the CHUCK.

Line 23 CLOSEs the HAND onto the component.

Line 24 Digital OUTput signal 1 is ON to open chuck.

Line 25 Sets a DELAY of 2 seconds.

Lines 26 and 27 MOVE the robot TO the CONVEYOR OUT position via the PARK position.

Line 28 OPENS the HAND to release the component.

Line 29 Sets a DELAY of 2 seconds.

Line 30 MOVEs the robot TO the CONVEYOR IN position.

Lines 31 and 32 CLOSEs the HAND onto the next component and waits for 2 seconds.

Lines 33 and 34 MOVEs the robot back TO the CHUCK via the PARK position.

Line 35 Digital OUTput signal 2 is ON to close the chuck.

Lines 36 and 37 sets a DELAY for 2 seconds before releasing the component.

Line 38 sets a DELAY for 2 seconds.

Line 39 MOVEs the robot TO the PARK position.

Line 40 adds 1 to the PART COUNT 1 register.

Line 42 if the Digital INput signal 3 is ON the loop terminates and the program proceeds with line 43. If the signal is off the program is repeated from line 20.

Line 44 displays on the operator's console the total number of components that have been machined.

Line 46 ENDs the program.

Problems

1 Explain how the following functions relate to the programming of a manipulator: (a) real world coordinates; (b) Cartesian coordinates; (c) tool centre point; (d) frame definition.

2 The following software functions are available in most robot controllers. Describe their operation and give examples of typical applications: (a) weave control; (b) mirror image and shift; (c) skip function; (d) subroutines; (e) registers.

3 Describe the procedure followed when planning a program for a robot to execute a pick-and-place task.

4 State three methods of programming industrial robots and for each outline any disadvantages that you consider relevant.

5 What methods of teaching a robot are suitable for the following applications? (a) Paint spraying; (b) Spot welding of car bodies; (c) Application of underseal; (d) Arc welding; (e) Machine tool loading.

6 List the keys that you would expect to find on an industrial robot teach pendant and briefly describe the function of each.

7 State the advantages of programming by means of off-line programming techniques.

8 What is the purpose of the robot controller's operating panel? List the functions that it performs.

9 Draw a flowchart and write a program for the following robot task.

A robot is used to load and unload two machine tools A and B. The parts arrive on an input conveyor and leave the machining cell on an output conveyor. The sequence is as follows:

(1) A part is picked up from the input conveyor on receipt of a signal that it has arrived.

(2) The part is loaded into machine A and machined. During this operation the robot moves to machine B and unloads a finished component onto the output conveyor.

(3) Machine A is then unloaded and the part is taken to machine B. While it is being machined the robot returns to its 'park' position and waits for an input signal from the input conveyor indicating that a part is present to be loaded into machine A.

(4) The sequence is repeated 100 times.

10 Rewrite the program in Problem 9 to include 'wait' signals when parts are grasped and released. Include a subroutine program which instructs the robot to stack components from a third conveyor onto the output conveyor if there are no parts present on the input conveyor for machining.

6 Performance specifications of industrial robots

6.1 Factors influencing the choice of a robot

Many factors influence the choice of a robot and the peripheral equipment for implementing a particular manufacturing task. The initial work study exercises and cost benefit studies should be carefully analysed to ensure that the appropriate robot is purchased. The following are key considerations.

Required accuracy and repeatability of the robot

It should be borne in mind that the better the repeatability the more the robot will cost and generally the working envelope will be smaller, i.e. the SCARA type. The required repeatability is really determined by the robot application. For example, robots being used for assembly tasks, welding and sealant applications require repeatability of the order of better than ± 0.2 mm, whereas for machine tool loading/unloading and palletizing any minor inaccuracies can be taken up by the gripper. This also applies to spray paint applications, where the nature of the spray pattern counteracts any of the errors in repeatablity.

Lifting capability

This may seem quite an obvious consideration, but a factor of safety should be included to allow for unforeseen loads, e.g. tight fits during assembly operations or the extra forces involved when attempting to assemble a defective part. It is important to remember that the robot load capacity includes the weight of the end-effector which in some cases can be considerable, as well as the maximum total weight of the workpiece. In some cases the recommended load capacity of the robot can be exceeded but this can only be achieved at the expense of the cycle time/speed.

Size of the working envelope

The working envelope should be large enough to accommodate all areas in which the robot needs to move in order to carry out its task efficiently. This spatial envelope is determined by the type of robot configuration, the number of axes and how the robot is mounted, e.g. overhead gantry, wall-mounted, floor-mounted or on tracks.

Number of axes required for end-effector orientation

There are generally two or three axes on most industrial robots' wrists, not including the gripper open/close. These can, however, be increased by incorporating further movements into the end-effector design, driven, for example, by stepper motors or pneumatic cylinders.

Controller software options

These are covered in Chapter 5 and include such options as incremental programming, shift function, rotation function, mirror imaging and software for welding and palletizing. If the robot is intended to be multifunctional in its application, then careful consideration should be given to the availability and cost of optional software.

Type and design of end-effectors

The end-effectors that are to be used on the robot, and how they function with the task to be performed, must be taken into consideration: also their association with peripheral devices for operations such as handling and feeding components or parts.

6.2 Robot performance testing

The performance characteristics of robots vary according to the configuration used and the manufacturer's build specification. Purchasers who have relied on the specifications outlined by the supplier may have to cope with discrepancies between actual and expected performance that are encountered during feasibility tests. A number of institutions, such as PERA, CRAG and IPA Stuttgart, have carried out comprehensive tests on many makes and types of robot as part of research activities: information and advice on the performance of different robots can be sought. It is possible, however, for the robot user to carry out his or her own evaluation of some characteristics in-house, using relatively simple equipment to establish, for example, the path accuracy with different payloads and speeds, perhaps to determine the deterioration in accuracy due to robot wear.

Robot characteristics that can be determined in-house include:

- Path/point accuracy and repeatability.
- Maximum working volume generated by the robot manipulator.
- Kinematic and static values such as accelerations and maximum speeds and the errors incurred due to arm and gripper loads when the robot is stationary.

The dynamic characteristics of a robot such as the load capacity, the elasticity, rigidity and weight of the arm and the maximum allowable gripping force are determined by the robot designer and cannot be influenced by the purchaser.

Path/point accuracy and repeatability

Before considering how accuracy and repeatability can be determined, the difference between these two properties in the context of robotics should be understood.

Repeatability is the measure of a robot's ability to go to a point or along a path exactly, time and time again under the same conditions: this point or path may not be identical to the one programmed. *Accuracy* is the measure of the difference between the 'true' point or path (the one programmed) and the actual point or path taken.

Figure 6.1 shows the difference between repeatability and accuracy. A robot has to position four parts in the centre of the circle: the result shown in Figure 6.1a indicates high accuracy and repeatability; Figure 6.1b shows the four parts close together but some distance away from circle centre – poor accuracy but high repeatability; Figure 6.1c indicates low repeatability and high accuracy, since the four parts are positioned close to the circle centre but in relation to each other there is considerable deviation.

To the programmer repeatability is the more important since inaccuracy can be reduced by reprogramming the point or path with the teach pendant to compensate for the error. This new point or path will then of course be repeated indefinitely.

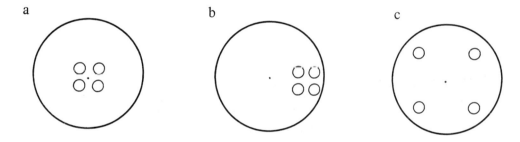

Figure 6.1 a High repeatability and accuracy
b High repeatability, low accuracy
c High accuracy, low repeatability

The method of programming has a considerable effect on the path accuracy of a robot. Figure 6.2 shows the result when an industrial robot was programmed to pass through five points; the path taken was plotted on paper by using a spring-loaded fine lead pencil held in the gripper. The points were entered as single point moves rather than linear path moves. It can be seen that each point was reached, but between two points no line was drawn. This was an articulated-arm robot, so its simplest path, in terms of arm movements, was a curved one – as is clearly indicated in some of the moves. In one case the pencil left the paper when there was a move in the Z direction. Thus the path taken between the points was unpredictable.

Figure 6.3 shows the path taken through the same points, but programmed in linear path control; this time the path is clearly a straight line, but there is a small curve at each of the points. This slight curvature is a function of the robot speed and the type of positional control programmed.

Figures 6.4 a and b give an indication of how the programmed speed affects path accuracy and how fine or coarse positional control relates to the robot path relative to the programmed position. As in the previous example, the robot was programmed to pass through the 5 points in a

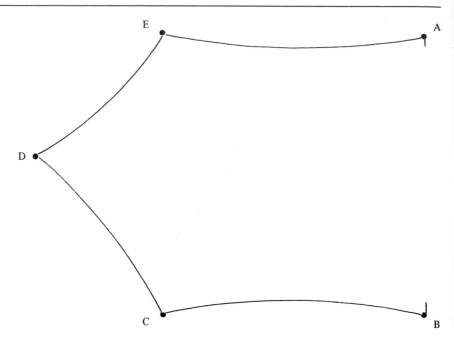

Figure 6.2 Path accuracy with single point moves

linear path, i.e. using Cartesian coordinates. The feedrates were 100 mm/s and 500 mm/s with coarse position control applied, i.e. no deceleration on approaching the point. It can be seen clearly that the path taken swept close to the point, but the path distance from the point was greater when the speed was increased to 500 mm/s. This can be improved by either reprogramming the path to take up this inaccuracy or by using fine positioning control, which is available on most robot controllers.

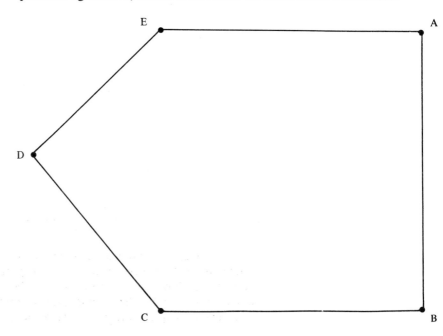

Figure 6.3 Path accuracy with linear path control

Figure 6.4 Linear path with
coarse position control
a Feedrate 100 mm/s
b Feedrate 500 mm/s

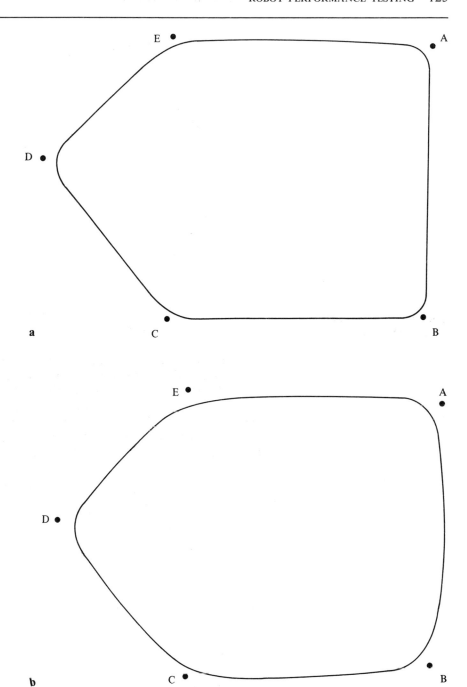

The same path was repeated for a speed of 100 mm/s with fine positional
control applied, i.e. deceleration when approaching the points. Figure 6.5
shows the path proximity to the point to be much closer than with coarse
control; predictably, the path deviation from each point would increase
with each subsequent increase of feedrate.

Figure 6.5 Linear path with fine position control, feedrate 100 mm/s

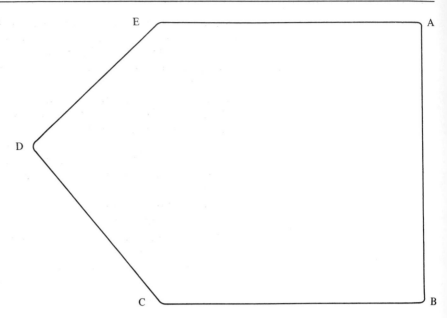

In the above tests the arm loading remained the same; similar tests could be repeated for different arm loadings, so establishing correlation between arm loading, speed and path/positional control accuracy. In developing a program it may be found that the robot will perform to the required accuracy at low speed but, when speeded up to production cycle timing, path errors become apparent. This can usually be overcome by inserting more points along the programmed path.

Positional repeatability

Positional repeatability can be measured accurately by programming the robot to touch onto dial test indicators situated around the working envelope. At different horizontal and vertical positions, the feedrate and arm loading can be varied to draw comparisons between their values and the relationship to the programmed position. The factors that affect accuracy for positioning and repeatability are:

- The feedrate and loading on the robot arm: with heavy loads, acceleration forces can cause considerable deflections.
- Backlash and play in chains, gears or other drive mechanisms including play in bearings: these can cause positional errors.
- Expansion or contraction of mechanical parts due to temperature variations.
- The resolution and precision of the controller's feedback system and of any sensing systems which are employed.

Determining the maximum working envelope generated by the robot arm

These are perhaps the easiest values to obtain and are associated with the maximum and minimum arm movements required to generate the working envelope of the robot. Most robot manuals give this information, but it may be useful to carry out this exercise in order to determine the envelope generated to the tip of the end-effector that is being used, since the manufacturer can only give the dimensions relative to the end-of-arm flange mounting.

Figure 6.6 shows the dimensions of the working envelope for an ASEA IRB 6/2 robot. These can be determined by moving all the limit switches to give the maximum range of joint position. Each axis is then moved to its maximum movement with the teach pendant, until the robot is stationary. The three axes on the wrist are then driven to give the maximum position, and the tip of the tool is measured relative to the floor and the centre line of the robot.

These positions are reached by driving the robot manually to them; when actually executing a program some of these points may not be attainable, due to mechanical and control constraints imposed by the software. The 'software envelope' can be established by inserting the maximum positions into the program by means of the teach pendant. The program is then executed using the single block mode, and at each position the X, Y and Z

Figure 6.6 Determining the maximum working envelope by driving the robot axes to their extreme positions (*Courtesy ABB Robotics Ltd*)

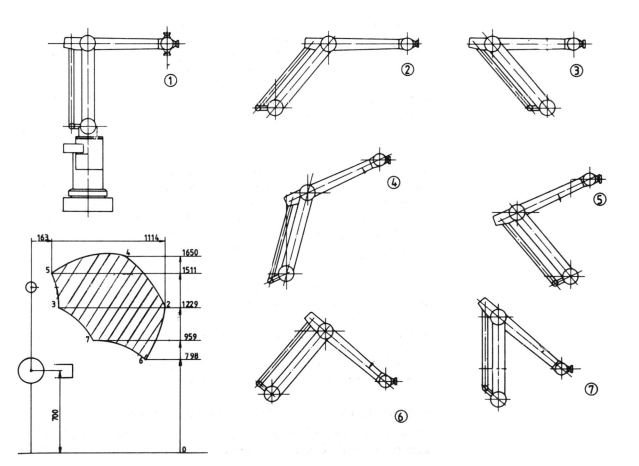

measurements are taken and compared with the end-effector position recorded manually. This test can be repeated for different feedrates and arm loading up to 100% of the manufacturer's recommended value.

Kinematic and static values

If the robot is being programmed in the workplace using the teach pendant, it is not generally necessary to measure the arm deflections that occur with various loads. Since the robot is taught the programmed points with the end-effector and load attached, any static deflections that are present within the arm are thus recorded as part of the robot's position. An indication of static deflection is more important when off-line programming, since it is the ideal position of the robot that is determined and entered into the robot program. The controller will not be aware that the arm has deflected, therefore no corrective action will be taken.

Arm deflection can be determined by positioning dial test indicators at different positions along the robot arm, as shown in Figure 6.7, with various loads applied at different arm reaches. Kinematic values such as acceleration and velocity can be determined by repeating a programmed cycle and recording the time taken to complete each cycle. This data will also be required to ascertain the production cycle time.

Some robots have the facility for the acceleration values to be entered for each move when programming. This is particularly useful if the task involves manipulation of delicate objects which may be damaged when subjected to high acceleration forces, or in preventing a part from slipping in the gripper. Small accelerometers attached at various positions on the arm will indicate the magnitude of acceleration or deceleration.

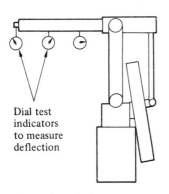

Dial test indicators to measure deflection

Figure 6.7 Determining the static deflection of a robot arm

Problems

1 What factors influence the choice of a robot for a particular manufacturing task?
2 Consult the literature of several industrial robot manufacturers and compare the performance specifications with regard to accuracy and repeatability, payload, axis speeds and working envelope size.
3 With the aid of sketches describe the difference between robot path accuracy and repeatability.
4 Show, with the use of sketches, how the path taken by a robot changes when:
   (a) the mode of operation is changed
   (b) the feedrate is altered
   (c) coarse control is replaced by fine control.
5 Describe the method used to determine the positional repeatability of an industrial robot.
6 What factors affect the accuracy for positioning and repeatability of a manipulator?
7 Describe, using sketches, the method for determining the maximum working envelope of an industrial robot.
8 Why, in some cases, does the software working envelope differ from the envelope generated by jogging each axis to its extremity of movement?
9 Explain how the static deviation of a robot arm can be measured.
10 State how the programmer can improve the accuracy with which the end-effector travels along a given path and around a particular point.

7 Robot cell safety considerations

7.1 Accidents involving robots

The first recorded fatal accident involving a robot occurred in Japan in 1978 when a maintenance engineer, repairing a robot-fed abrasive machine, was pushed by the robot from behind and crushed. A second fatality occurred when a Japanese worker jumped onto a conveyor belt to remove a bulldozer axle that was being transported to a machine tool for drilling: a robot was used for the machine tool loading operation, and a micro-switch was actuated which resulted in the robot crushing his back.

Up to 1985 five fatal accidents in Japan concerned industrial robots. In each case the accident was caused by the operator/worker being unaware either of the robot's movements or of the wait commands within the robot program. What may appear to be a known and familiar part of its cycle may deceive even the programmer who initiated the robot's sequence of operations. If the robot can confuse its programmer, consider the dangers faced by those who are trained only to service, supply or merely support the manufacturing process that the robot is engaged upon.

The main conclusions of various surveys conducted on accidents involving industrial robots are that the majority of accidents have occurred:

- During programming and the initial running of the robot sequence
- During the adjustment/maintenance of peripheral devices
- As a result of inadequate provision or faulty installation of safety equipment

The most important consideration when designing a robot cell or installation is the safety of the programmer, operator or maintenance engineer who is working in or around the robot operating envelope. Second to that is the protection of costly equipment, including the robot itself.

The substantial proportion of operator deaths in the fairly low number of accidents that have been experienced with robots is significant when compared with the low number of deaths occurring in the very numerous accidents involving dedicated machine tools. Robots therefore demand very special consideration, requiring specialized safety systems if the shop floor is to remain safe.

7.2 Factors affecting robot safety measures

The criteria for the design of a safe robot cell or installation differ from the safety requirements and layout for a conventional machine tool such as a lathe or machining centre. Robot operations have the following characteristics.

1   The working envelope can be large, depending on the type of robot being installed; a typical arm radius is 1.6 m with a sweep angle of 300° some 1.5 m above floor level. It can be difficult to visualize the working area since the 5 or 6 axes in synchronized movement generate a complicated three-dimensional trajectory in space.
2   Speeds can be high, of the order of 2–3 m/s, slew rates of 90°/s with payloads of 10–200 kg at maximum arm reach.
3   On point-to-point programming the robot will follow the shortest route between two programmed points; because each axis operates independently, it is difficult to predict the trajectory that the robot will take.
4   The programmer/operator has to be within the operating envelope when creating the program.
5   The robot may be stationary at unpredictable times during the programmed sequence, e.g. it may be waiting for an input signal from a peripheral device to say that a part is present.
6   More than one robot may be working within conflicting envelopes, e.g. spot-welding of car bodies along a line or spray-painting of body shells in the automobile industry.
7   Because of the diverse applications of robots, access to materials, parts and peripheral devices such as conveyors, bowl feeders and tool-change stations must be provided.

7.3 Safety features built into industrial robots

All industrial robots have a number of safety features built into their design to protect the working parts and to ensure safety of the operator when creating a robot sequence, editing an existing robot program, or executing a program in automatic mode. These generally fall into the following main functions:

1   If, on start-up or during the cycle operation, the robot moves beyond the designated axis travel the robot will go directly into an emergency stop or 'hold' condition. This does not necessarily stop the peripheral equipment, which may continue to send a signal for the robot to execute a particular operation. The program is not violated: after the relevant axis has been jogged back within its operating limits, the program will continue to be executed from that part of the program preceding the over-travel. On some older robots it may be necessary to move the robot back to its home or 'park' position and then re-run the entire sequence; this has obvious disadvantages as it may be necessary to reset internal and external control signals.
2   The robot will go into emergency stop condition if there is any drive or control malfunction. This could include failure of the control unit or drive motors through overheating, runaway of the servo-motors, loss of power supply or erroneous data in the memory.

3 A dead man's handle is often part of the teach pendant (see Figure 7.1). This can be either a two-way or a three-way device; the latter has the advantage that the robot will stop if the operator either tightens or releases the grip. Usually incorporated on teach pendants are a 'hold' key and an emergency stop button. 'Hold' will cause the robot to stop immediately: application of the brakes causes deceleration to each axis and the gripper defaults to the closed position, so preventing the workpiece or component from being dropped. When the reason for holding is rectified, the robot system is actuated by pressing the 'hold' key again. This enables the program teaching to continue or initiates a return to the automatic mode. A teach pendant incorporating these features is shown in Figure 7.2. These functions are also repeated on the operator's panel situated outside the robot cell.

Figure 7.1 Fanuc teach pendant showing the two-way dead man's handle and the enable key (*Courtesy Bournemouth and Poole College of Further Education*)

Figure 7.2 Teach pendant for Fanuc S100 robot showing the 'hold' key, emergency stop button and teach pendant enable key (*Courtesy Bournemouth and Poole College of Further Education*)

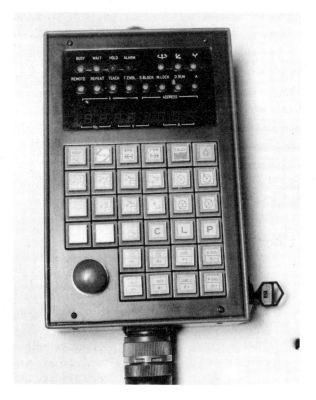

Actuating the emergency stop applies the brakes immediately and shuts down all peripheral equipment. In the earlier type of robot this can cause problems in re-starting, as the controller may not know the existing physical position of the robot or the relevant point in the program. Damage may therefore be caused to the robot or other equipment when continuing the program, as its path will be unpredictable. It may be necessary to return to the zero position by using the jog axis keys to withdraw carefully from the point of impact or malfunction.

4 The most vulnerable part of the robot is the end-of-arm tooling as this is the part that is in contact with external objects such as the

workpiece/component and the loading and unloading equipment. There are a number of ways in which the tooling and consequently the arm can be protected; these include breakaway torches for welding, wrist breakers that snap in two when too much pressure is exerted on the tool, and also proximity probes and tactile bars that detect the presence of an object. Some of these devices operate by means of a micro-switch positioned at the end of the arm which is actuated on collision, sending a signal to the 'hold' or emergency stop function in the robot controller.

7.4 Manufacturers' guidelines for robot safety

All major robot manufacturers issue guidelines regarding the installation of their robots and recommend safety precautions. These can be summarized as follows.

- Ensure that the robot's operating space and that of service equipment is secured from interruption by personal and/or transportation vehicles such as forklift trucks and AGVs when in operation.
- Provide an adequate number of emergency stop buttons in and around the robot system.
- When working within the robot's envelope, whether programming or carrying out maintenance work or inspection, always ensure that there is an escape path: never turn your back to the robot.
- The operator in the robot cell has priority over all others since he has the teach pendant and knows which moves the robot arm is going to make. Any operation of the controller or interface equipment must be supervised by this operator.

Special precautions for routine maintenance and inspection

1 The person entering the robot working area must have the teach pendant, so that no one else can actuate the robot. This can be incorporated into the security system, with access only possible when the dead man's handle is pressed. This is particularly important if 'power on' is required for maintenance, as the engineer, if necessary, can use the emergency stop button on the teach pendant.
2 If power, either electrical or pneumatic, is not required, it should be isolated at source before the robot cell is entered.
3 Always assume that the robot may be about to be actuated: many accidents have occurred when the robot started to move after being stationary for a period.

7.5 Safety aspects of the robot system

Cables

Accidents such as tripping can result from loose cables to/from the control cabinet and robot, or to peripheral equipment. These should be either housed in box sections set into the floor, or fixed with a recommended covering. This also protects the cables from wear and tear and work hazards such as weld spatter, chemicals, and moving and rotating parts.

Grippers

The prime function of the gripper is to grasp the workpiece continually with the required pressure, and to hold it until otherwise instructed. In the event of a power failure the gripper should default to the closed position to ensure that the part is not dropped or thrown. This can be achieved by using bistable valves with a pressure switch to maintain a constant force and dual circuitry to compensate for a power failure.

Speed control

When an industrial robot is programmed from the teach pendant the operating velocity defaults to around 25% of the actual programmed speed. Once the program has been taught and proven, the feedrate can be globally changed within the program on the operator's control panel or overridden by using the teach pendant feedrate % key. This procedure is only recommended when operating from outside the robot cell. It is possible on some robots to have a speed override when operating in repeat or automatic mode. This has the effect that, when the safety barrier is broken, the robot defaults to a preset percentage of the programmed feedrate, say 30%, instead of going into a 'hold' or emergency stop mode. This allows a person to enter the cell to make adjustments or carry out inspection while the robot is in operation, without being subjected to the danger of the robot moving at full production speed.

Braking system

It is essential to have brakes on each joint of an industrial robot to maintain its position in space whenever it is turned off, receives a 'hold' command, or in the event of a power failure. As the drive systems employed (e.g. recirculating ball screws, harmonic drives) are virtually frictionless, the robot joints would otherwise fold under the weight of the end-effector load and the loads exerted by each axis of the arm, resulting in loss of position and possible damage to the robot. Restarting the robot is obviously also made much easier.

Since the brake assembly fitted to each joint needs to be compact and reliable, calliper brakes, similar to those used on cars, are used. These are operated in a fail-safe mode, i.e. the brakes are on when there is no power supply to them. The DC drive motor for each axis also has to be stopped and held in position. This is achieved by reversing the current flow through the armature, producing a counter e.m.f. which stops the motor.

7.6 Safety barriers and other devices

In any robotic system there are a number of elements that can cause accidents as a result of malfunction or failure. These include mechanical elements of the robot, input/output signals from controllers, interfacing units and the mistakes or ignorance of persons directly within the vicinity of the robot cell. As stated previously, many safety factors have to be taken into consideration if the robot is to perform at its designed production rates with the minimum of danger. This can only be achieved if all parts of the system are operating reliably and safely.

Safety barriers are designed to prevent personnel from entering the cell when the robot is in automatic mode. There are many different ways in which they can be formed. Perhaps the simplest is the use of peripheral equipment to form a natural barrier: conveyor systems, machine tools or pallets positioned around the extremity of the robot movement. This does not, however, prevent a person from deliberately climbing over or from throwing objects into the cell when the robot is moving. In general, the following physical safeguards are employed.

Fixed wire mesh guarding

This is used in many industrial applications since it is cheap, effective and easily installed. Figure 7.3 shows a typical installation of wire mesh guarding. For most installations this type of guarding fence should be a minimum of 2 m high with a mesh of maximum 50 mm square to prevent hands from entering the robot area. This type of peripheral fixed fencing does not necessarily prevent operatives from climbing over, under or even through 'chicken wire' barriers. When using this type of barrier there are various considerations to take into account:

Figure 7.3 Fixed wire mesh guarding around a palletizing cell (*Courtesy GMFANUC Robotics (UK) Ltd*)

1 It may be difficult to transport parts of components into the robot cell, e.g. car bodies or engine sub-frames for welding, front and rear windscreens for assembly into the car aperture.

2 Sliding doors and/or work loading hatches are needed to provide adequate access for equipment, materials and maintenance requirements.

3 Door safety interlocking systems are required to stop or hold the robot whenever the doors or hatches are accessed. There are many different types of interlocking switch available, varying from magnetic to microswitch operation.

    This barrier type of protection to a working area can itself be a danger to personnel: if the operator of a robot installation were intentionally or otherwise to lock himself, or others, inside the robot work area during the robot operating cycle, the safety fencing designed to keep people out would now keep them captive; the risk is then of being crushed or pinned between the robot and the barrier fence.

4 Wire mesh is difficult to see through, particularly for long periods of time. If the operator is required to implement many visual checks it may be necessary to replace the mesh with toughened Perspex windows.

5 Fixed fencing does not provide the ultimate in ensuring the safety of workers. To be commercially viable, sophisticated and expensive robot installations must function to their fullest production capacity and to achieve this they must be provided with equally sophisticated and efficient safety systems geared to this objective.

Pressure pads or matting

These pads, of varying size, have either tiny air tubes or fibre optic sensors sandwiched within the mat. When foot pressure is exerted on the mat the air back-pressure, or change in light intensity, is detected by the controller and the robot is put into a 'hold' or 'stop' condition. The operation is very fast and is reliable since if the tube or sensor is damaged, broken or short-circuited the control will fail to safety.

    These devices operate over long distances, e.g. 1 km, and can be used in hazardous areas. However, in some applications a protective covering may be required, as in the case of welding where weld spatter may burn the matting. Resting heavy objects on the pads is not advised; this may cause some inconvenience when positioning welding fixtures, conveyors and other robot ancillary equipment.

Infra-red light curtains

Light curtains as safety barriers for robot installations are becoming increasingly popular due to their flexibility in use and reliability. An infra-red curtain comprises a combined infra-red transmitter/receiver unit, mounted vertically or horizontally, a microprocessor-controlled safety monitoring system and output storage providing an interface to the robot controller.

Robot operation is initiated from the operator's control panel which is mounted outside the protected area. Provided that the light curtain is not interrupted, the robot sequence will begin; however, any intrusion on the light beam will generate an electrical 'stop' signal. This places the robot under direct supervision of the light curtain monitoring system. If the reset button is not depressed before expiry of a predetermined time period, a 'hardware' stop will follow; this shuts off all power to the robot and ancillary equipment. This type of system prevents the total shutdown of the robot should the light curtain be accidentally interrupted.

Three examples of light curtain installations are shown in Figures 7.4, 7.5 and 7.6.

Advantages of light curtains

- A versatile, simple and reliable system which eliminates the use of traditional heavy guarding, thus providing easy cell access for materials and component parts.
- When extending the cell area or changing the cell layout, the transmitters and receivers can easily be removed and repositioned allowing simple redefinition of the protected area.
- The light curtain can be bent through any angle by using mirrors; this enables a complete cell to be protected with only one

Figure 7.4 Lightguards light curtains installed at the ASEA training school, Sweden (*Courtesy Lightguards Ltd*)

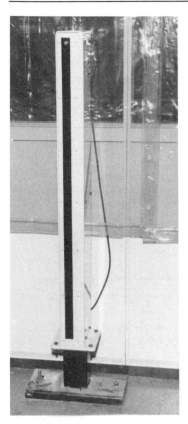

transmitter/receiver. They can be used either vertically or horizontally; the latter gives the advantage that the complete base area is monitored, thus eliminating the need for a safety interlock.

- In the case of automated production where the robot is fed by AGVs, the light curtain can be disabled for a set period of time on receiving a mute signal from the AGV when it is arriving.
- The area protected by the light curtain can be indicated simply by using coloured tape on the floor.

Figure 7.5 Lightguards infrared light curtain providing a safety barrier for a robot cell (*Courtesy Bournemouth and Poole College of Further Education*)

Figure 7.6 (*right*) A Lightguard protection barrier (*Courtesy Lightguards Ltd*)

Electromagnetic field safety barriers

This type of barrier is similar to light curtains in operation except that an electromagnetic field is generated instead of an infra-red curtain.

A coupler is mounted onto a handrail around the robot cell, creating a sensing perimeter guard which prevents intrusion by the operator or unauthorized personnel. The system consists of three basic components: control unit, coupler and interconnecting cable. The control unit generates an electrical signal which is supplied to the coupler through the cable. The coupler is electrically connected to the sensor, and causes an electromagnetic field to be set up throughout its length. Interruption of the field will cause an electrical signal to be transferred to the control unit by the coupler. The relay in the control unit de-energizes, causing the desired control action to be initiated.

The sensor can also be mounted on a robot arm to prevent collision with other robots or peripheral devices.

7.7 Some typical installations of safety barriers in robot cells

Robot welding station with manual feeding

Figure 7.7 shows the arrangement of a robot welding station comprising a welding robot and a rotary table to allow components to be fed into the cell and inverted, so enabling the robot to reach all of the necessary welding points.

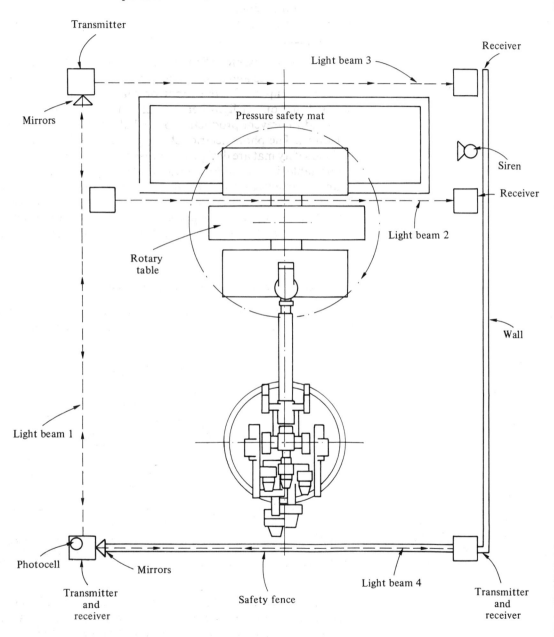

Figure 7.7 Safety barriers for robot welding station

The operating cycle is as follows:

1 The operator removes a welded workpiece from the rotary table and clamps the next one in position.
2 The operator gives the signal that the workpiece is ready for welding.
3 When the robot is ready for the workpiece, it signals the rotary table to move into position.
4 The robot welds the new workpiece and the operator removes the welded workpiece from the table and mounts the next workpiece to be welded.

The cell comprises four light beams, a pressure safety mat, a photoelectric cell, a safety fence and a siren. Light beams 1, 2 and 4 protect the operator from being injured by the robot or rotary table. The photoelectric cell situated in front of light beam 1 and the fence below light beam 4 are designed to prevent production being halted inadvertently by intrusion of personnel. The photoelectric cell triggers an audible alarm. Light beam 3 and the safety mat are designed to prevent the operator being injured by the rotary table, i.e. the table will not rotate when the operator is in that area. Beam 2 is inactive only when the safety mat and beam 3 are not triggered. The dead man's handle on the teach pendant can be wired into beam 1 circuit so that when it is depressed the beam is switched off, allowing the operator to enter and leave the zone without interupting the welding sequence.

Robot spot-welding on a car production line

Figure 7.8 shows sections of a car production line separated into various operating zones. To enter a zone, a person must first open a gate,

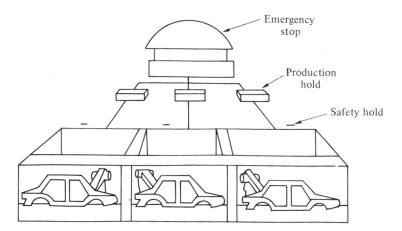

Figure 7.8 Robots spot welding on a car production line separated into distinct operating zones (*Courtesy Lightguards Ltd*)

whereupon a safety hold is initiated by limit switches. It is not possible to walk from one operating zone to another. Thus, it is possible to halt production and carry out work in one operating zone while production continues in the other zones. The same interface on each robot is used for connecting the production hold, safety hold and emergency stop, which

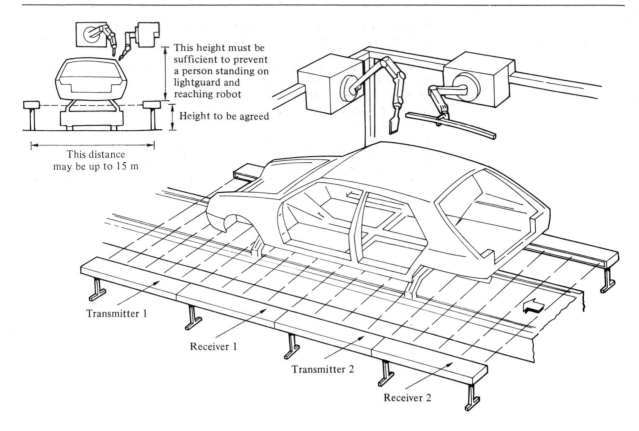

This height must be sufficient to prevent a person standing on lightguard and reaching robot

Height to be agreed

This distance may be up to 15 m

Transmitter 1

Receiver 1

Transmitter 2

Receiver 2

Figure 7.9 Diagrammatic representation of light curtains covering the complete track of a car spot welding station (*Courtesy Lightguards Ltd*)

means that all potentially hazardous machining operations are stopped securely in each case. For some minor pneumatically controlled functions, only programmed holds are used, since the use of other stop/hold functions would create major problems for restarting the production line. Infra-red transmitter/receivers can be applied similarly to the welding of car bodies, shown in Figure 7.9. The light curtain covers the complete area of the 'track'; if the curtain is broken by an intruder the robots go into a 'hold' function and the welding current is switched off. If the curtain is not reset within a specified time, the system goes into a hardware stop.

Problems

1 What special features of a robot operation affect safety criteria for a robot cell, as compared with a conventional machine tool such as a lathe or a milling machine?
2 Explain the function and operation of the safety features incorporated in a robot teach pendant.
3 Explain the difference between the 'emergency stop' and 'hold' functions. When would each be used?
4 List the precautions that the operator should take when programming within the robot's working envelope.
5 Explain the safety features inherent in the design of a gripper.
6 Why do robots have brakes? How do they operate?

7 Explain the advantages and limitations of use of the following safety measures: (a) fixed wire mesh guarding; (b) pressure pads or matting; (c) infra-red light curtains.

8 Describe the operation of an infra-red light curtain. What are the advantages of this type of barrier over conventional guarding?

9 Describe and illustrate the layout of a robot cell for welding, showing the following safety devices:
    (1) Light curtains
    (2) Pressure pads
    (3) Fencing with door interlocking to the robot controller.

10 Discuss the safety precautions that must be taken when several robots are working in close proximity to each other, e.g. spot welding on a car body line.

8 Cost justification of robots

8.1 Job analysis

When considering the implementation of either a single robot application, such as a simple pick-and-place task, or an installation which is using groups of robots, as in the case of a full flexible manufacturing system, many factors must be taken into consideration.

From the initial concept of robotization it must be questioned whether the purchase of a robot or robots is a viable proposition in terms of efficiency, economics and practicability. In order to carry out the initial evaluation it is necessary to set up a small robotics team consisting of capable and responsible people. They need to research and to analyse all the information gathered. In order to achieve positive feedback the team needs to be balanced and neutral, as a strong pro-robot team is not necessarily going to look at the advantages and disadvantages of robotization with an open mind. Various aspects need to be studied and discussed.

Analysis of the existing work system

From the very beginning this stage of the study is of vital importance and, depending on the correlated information, could become the benchmark against which to assess the merits of implementing robotization. Initially, it is necessary to identify the company's current and future needs and to establish the changes that are likely to take place in the future. The following considerations are relevant to future commercial development:

- If it is required to continue competing in a particular market, what needs to be done to hold on to, or to increase, the company's market share?
- What is required to improve the perceived quality and reliability of the product range?
- What are future market requirements likely to be and what is the expected rate of change?
- Can the costs be reduced in such areas as inventory, materials, capital equipment, reworking of defective components, manpower?

From management control and the relevant documentation to shop floor operations, all activities need to be studied carefully, bearing in mind that a target is being aimed for and also a task is being set. Such a task for a particular company might be to increase the production capacity, currently running at, say, 30% below that of customer demand – the significant factor being that each machine is perhaps only 50% utilized when operated manually. Could the introduction of industrial robots give a significant increase to around 80% utilization of all machines?

The current working system should be examined with robots in mind. One of the main exercises here is to eliminate the possibility of bottlenecks occurring in the manufacturing process – which can happen no matter how many robots are installed. This will form the basis for identifying where robots can be used effectively.

At this stage a sorting-out process is required: all possible robot applications should be listed and investigated as to their possible effectiveness for performing the particular task. The arrangement and dimensions of working areas need to be taken into account. The siting of relative peripheral devices should be studied carefully, including whether any need to be within the robot cell.

Throughout this process the team should continue to question whether the introduction of industrial robots is the correct step to take. Comparisons should be made bearing in mind that the robot has two main competitors, people and hard automation.

Robots, unlike people, cannot handle crisis situations, so robot applications are somewhat limited even with the use of sophisticated visual and tactile sensing. However, although they have this lack of intelligent response, robots have distinct advantages over hard automation.

- Hard automation tends to be inflexible, generally being dedicated to a particular product or task.
- It can well outlive the produce for which it was developed and equally the equipment can become obsolete early in the product life.
- It necessitates holding large inventories of raw materials, and of finished parts when production runs are extended.
- Such plant presents difficulties in commissioning and high maintenance costs are also a consideration.

By contrast, a robot is:

- Quick to commission
- Reprogrammable for different tasks
- Easy to maintain
- Cheaper to design, because the difficult elements are pre-engineered

8.2 Implementing robotization

The decision to implement robotization involves considerable changes within an organization. Linked with the identification of an application is the identification of the changes that need to be made across the company structure, involving both management and shop floor workers, to allow for smooth implementation of robotization. The introduction of robots into the workplace is likely to have a significant impact not only on the worker but also on management and the organizational unit.

Impact on the worker

One of the causes of worry when robotization is being considered is the possible displacement of workers. This can mean a lost job, but equally it can mean employment on a task that may be either mentally or physically more demanding. Although the new workplace may be a much healthier environment and the working conditions may be better, the position could well carry more responsibility, e.g. where a technician moves into a supervisory role. The word 'bumped' encountered in this context expresses the workers' anxiety about one of the following types of displacement:

Job loss
Less desirable work
Less desirable shift
Lower pay

Careful planning can minimize these problems, but they have to be given serious consideration, not least because a high rate of lay-off that can be attributed to new technology can jeopardize industrial relations.

Impact on management

The introduction of robots will always create new job opportunities. Production workers, for instance, may take on responsibility for the operation of the robot and promotion to management could and does happen following the organization of a robot team. The main consideration is that management should be fully aware of the individual skills of workers so that jobs can be matched to personnel, with whatever retraining is necessary. Management also needs to assess the need for support such as maintenance, engineering and quality control personnel.

Work scheduling must be carefully planned. Retraining needs must be identified and provided for, possibly with a roboticist brought in to implement an on-site training program.

Impact on the organizational unit

Questions of organizational structure are often regarded as subordinate to questions of strategy and human resources. However, the introduction of the robot in the factory will have direct consequences for the organization itself. The effectiveness of the robot installation will be determined partly by the success of appropriate modification of the organizational structure.

8.3 Identification of the capital and recurrent cost of investment

Most companies, large or small, have some form of cost justification associated with the purchase of new capital equipment. This justification procedure can be anything from a few hand-written notes to a complete and detailed cost–benefit analysis. The purpose of all justifications is the same.

Since a company's resources are not unlimited, it is necessary to decide how available resources will be applied to various corporate needs. In most cases, this is done by developing budgets for annual and longer-term expenditure. Each company has its own formula for arriving at the optimum balance. Gut feeling, management philosophy, competition and predicted market behaviour all contribute to a series of decisions that parcel out the available resources of a company. These decisions have a great deal to do with success or failure of the company.

Capital expenditure appropriation

This represents the funds available for new capital expenditure during the budget period. Once the amount of money available for capital equipment has been determined, it is simply a matter of looking at the requirements for new equipment and comparing the two. If the cost of new equipment requested is less than the money appropriated, everything should balance up nicely, but this seldom happens – more usually each department sends in a 'wish list' and the total is considerably more than the capital expenditure appropriation.

If we assume that investment in robotization is being considered, a three-part analysis is required which will allow us to determine the profitability of the total initial investment:

1 Determine the investment required.
2 Measure the effect of the investment on operations, costs and profitability.
3 Calculate the return in relation to the required investment.

The investment required is arrived at by adding all the final costings together. However, where a feasibility study is being carried out for the first time and the company lacks experience and knowledge of the robot industry, a safety factor of around 20% should be incorporated in the final cost. This also provides a contingency fund for unforeseen expenditure.

Having determined the total investment need for the robot application, the next step is to correlate the data in such a way that a chart may be drawn up to show the intended profit. This can be done on an hourly, daily, weekly, monthly, half-yearly or yearly basis, as long as the formula is used consistently throughout the exercise.

To calculate the value of increased production capacity resulting from the installation of an industrial robot, the sales value of a single part is taken and the direct materials cost associated with that part subtracted from it. The remainder is multiplied by the number of additional parts that can be produced over a specified period as a result of the robot installation. The result of this multiplication is the new production effect.

Calculating the payback time on the basis of time

When capital equipment has been purchased a major consideration is the payback time. The following example demonstrates a simple method of determining the payback period of a robot based on time.

Example

	£
Cost of robot	27 000
Labour replacement cost	10 000
Maintenance cost for one operative	1 500
Maintenance cost for two operatives	2 500

(i) *For one operative*

The payback time is the cost of the robot divided by the labour cost minus the maintenance cost, i.e.

$$P = \frac{I}{L-E} \qquad (1)$$

where P is the payback period in years
I is the total capital investment in the robot and accessories
L is the total cost of labour replaced by the robot
E is the annual cost of maintaining the robot

Substituting the above values into (1),

$$P = \frac{27K}{10K - 1.5K}$$

$$P = \frac{27K}{8.5K}$$

$$P = 3.2 \text{ years}$$

(ii) *For two operatives*

Substituting the values for two operatives into (1),

$$P = \frac{27K}{20K - 2.5K}$$

(note that the labour replacement cost is doubled for two operatives)

$$P = \frac{27K}{17.5K}$$

$$P = 1.54 \text{ years}$$

Calculating the payback time based on production rates

The above method can be taken further and used in a production rate payback calculation.

Example

	£
Cost of robot	27 000
Labour replacement cost, based on £5 per hour, 8 hours/day for 250 days	10 000
Annual maintenance cost	1 500
Depreciation calculated at 15% of the total cost of the machinery at £100 000	15 000
Production rate coefficient (comparison with human operator)	±20%

The payback time is the total capital investment in the robot divided by the annual labour saving cost, minus the annual maintenance costs, $\pm$ the production rate coefficient, times the annual labour saving cost plus the annual depreciation costs:

$$P = \frac{I}{L - E \pm Q(L + Z)} \qquad (2)$$

where P is the payback period in years
I is the total capital investment in the robot
L is the annual labour saving cost
E is the annual maintenance cost
Z is the annual depreciation cost of associated machinery
Q is the production rate coefficient – this is either a plus or minus quantity

Substituting the values into (2),

$$P = \frac{27K}{10K - 1.5K + 20\%(10K + 15K)}$$

i.e.

$$P = \frac{27K}{8.5K + 20\% \times 25K}$$

$$P = \frac{27K}{8.5K + 5K}$$

$$P = \frac{27K}{13.5K}$$

$$P = 2 \text{ years}$$

It should be noted that this formula shows the production rate coefficient $Q = +20\%$ with the identification of the capital and recurrent costs of the investment.

Calculation of the effect on profits

The calculated effect of a robot installation on profit levels forms the basis for target figures for profits during the first year from installation and possibly beyond.

Production capacity effects

The production capacity effect is directly related to increased capacity, i.e. to an increase in the value of sales after direct materials costs have been deducted. The total capacity effect is the total return per year.

Operating costs

These include:

> Labour
> Indirect maintenance
> Programming
> Depreciation
> Robot tooling, installation and engineering
> Other operating costs

Savings

Direct savings following the robot installation contribute to a marked reduction in overhead expenditure in the following areas:

> Direct hourly wages
> Social expenditure
> Production costs
> Materials consumption
> Quality control
> Internal transportation
> Energy consumption

Examples include:

- Reduced reject rates
- Savings in raw materials
- Safety and health program savings

8.4 Planning a robot installation

The schedule

To establish a practicable program for the introduction of the robot, a simple but effective schedule of work must be set up. This schedule should cover planning from installation to programming and commissioning. It needs to be closely administered, to the point where, no matter how many people are involved, they all know exactly what has to be done and when. The program should be planned to cause the least possible hindrance to work in the area surrounding the robot cell installation. Planning for the introduction of the robot therefore needs to start as soon as the decision to install has been made.

Typical schedule for installation

1 Break down the existing system.
2 Prepare the robot site.
3 Rearrange the existing production facilities.
4 Revise the production control and planning systems to maximize the robot utility and to achieve the targets established for the system.
5 Create buffer stocks and arrange the necessary storage facilities.
6 Organize training, retraining, redeployment and, if necessary, redundancy.
7 Negotiate the necessary trade union agreements.
8 Design and build new jigs and fixtures.
9 Review existing quality control procedures.

These activities will not necessarily take place in this exact sequence but all are equally important and deserve careful attention.

If a thorough and sound feasibility study has not been carried out, this kind of exercise often highlights areas of manufacture other than those originally identified as requiring action, as well as shortcomings and possible improvements in the existing system. To automate an inefficient manual system may simply be to emphasize and perpetuate its inefficiency.

Installation

After all the hours of analysing, planning and preparing, this is the stage when the hard work begins, assuming that the robot arrives according to plan.

The schedule should provide for adequate time for the testing of all equipment when the robot is delivered. There is no point in accepting substandard performance from any equipment, as this will only build up problems at a later stage. In making this point, it is necessary to recognize that in the process of installation timescales tend to concertina towards the testing time: there is always the temptation to let something go in order to keep to a schedule. Critical checks include:

- Do the conveyors maintain their speed under load and can the speed be adjusted accurately?
- Do work handlers or grippers work correctly?
- Do the jigs and fixtures operate correctly?

No matter how small the piece of equipment, be sure of the performance before assembling the system. Problems may occur during installation; strict control must be kept on the flow of installation and each problem, as it occurs, must be rectified before going on.

Commissioning

This is the trial period during which programs are debugged as necessary, so in theory no cost is incurred apart from the cost of workpieces scrapped in proving the system. The only other cost that may occur is if extra work is required that is not on the original quotation for the job.

The length of time for commissioning will obviously vary in view of the number of possible machines and peripheral devices in any one cell. However, to keep this time to a minimum it is not uncommon for the supplier to build the cell at his premises and run a trial batch of work through the system, with the purchaser present, to see that all is working as it should. This last run is used to eliminate any faults in the system before it is dismantled and delivered to the purchaser. There are bound to be teething problems during the first few months and the purchasers will invariably need all the help that they can get.

Commissioning is always a useful time for the prospective maintenance engineers and programmers to watch how the cell is installed and to ask any pertinent questions about their particular parts of the system.

8.5 Maintenance

It is normal practice for the supplier to provide a warranty with the equipment. This usually covers the cost of parts and labour for one year, after which a maintenance contract is entered into at a rate previously set with the costing of the equipment and the final quotation. Typical figures quoted for maintenance intervals give the life cycle of connection cables as around 10 000 hours, and failure rate (or down time) = 0.03, which represents one failure every 33 months; this figure is being improved as recent studies indicate robot manufacturers giving a value approaching 0.02, or one failure every 50 months.

The natural progression in design and manufacture of the industrial robot has shown a marked increase in quality and reliability, working towards the totally maintenance-free robot. Drive motors are an example: the standard has improved from the use of stepper motors, to brushed DC servo motors, to the AC servo motor that we see in use today. This uses a digital encoder for instant feedback of position and has absolute positioning capacity, which means that the robot does not have to be referenced every time it is powered up.

Normally a maintenance engineer would have to make regular checks, from basic monthly greasing checks to quarterly, half-yearly and yearly checks which get more demanding as time goes on. A lot of this work is now eliminated by the use of direct drive motors that are totally maintenance-free. Spare parts holding is confined to basic items such as fuses, the odd cable and wrist joints.

8.6 Obsolescence

It may seem strange to talk about obsolescence in such a new technology, but it is inevitable. However, the industrial robot is built to last and is a well engineered piece of equipment. A rough estimate would be that a company would be looking to change a robot after 5–7 years, but this greatly depends on the number of shifts the robot is being worked, the type of work and the working environment.

There could well be a fine line between a robot being obsolescent and being out of date. Progress in the development of software and hardware may well contribute to this judgement, but if the robot is carrying out its task in a satisfactory manner and maintains the flexibility to cope with the ever-increasing variety and demands of production, who is to say that the machine has had its day? It is only when spares run out, the performance of the robot wanes and it cannot maintain its repeatability that the question of obsolescence arises.

8.7 Life cycle costing

The company accountants will be monitoring the capital cost of the robot equipment with a view to maintaining the target time for the payback period, especially if this is the first capital expenditure of its kind. The following questions will help with the life cycle costing of the equipment purchased:

- How soon will the robot realize its full working potential to the point where a profit is being made?
- What percentage of increased throughput can be expected within the first year?
- What is the measured effect of the investment on operations and profitability?

Through the fiscal year the sequence of events should be tightly monitored for this costing exercise. Taking into account the projected payback time and on the basis of the life expectancy of the robot, the accountant should be able to use these figures for the life cycle costing of the equipment. As has already been mentioned, this costing exercise will depend on what work the robot is carrying out and how many shifts the robot is working.

Problems

1 State four factors that should be taken into consideration when planning a company's future commercial development.
2 List the disadvantages of using 'hard automation'.
3 Outline the possible effects on the work force when industrial robots are introduced on the shop floor.
4 How do small and large companies arrive at cost justification plans for capital expenditure?
5 Describe how a company's capital expenditure appropriation can be broken down and outline the areas for analysis.
6 Explain a simple payback method that would be used for a robot installation and state how the total payback time is derived.
7 Outline the factors taken into consideration when determining the payback method for a robot investment on the basis of production rates. Explain why this method of calculation differs from that used in Problem 6.
8 What factors contribute to the operating costs of a robot installation?
9 As a result of a feasibility study your company has decided to install a robot. What requirements do you need to take into account, when drawing up an installation schedule, to ensure that the installation is successful?
10 Explain the difference between the terms 'installation' and 'commissioning' in relation to a robot work cell.

9 Robot applications

The basic types of robot have already been discussed and their variety gives an indication of their flexibility in terms of application. The choice of a robot for a particular industrial application is limited by considerations such as size, configuration, payload capacity and number of axes.

The industrial robot is an ideal tool for use in hazardous environments: the operator can be moved to a safe area where his skills are used to maintain the robot cell. For instance, in welding the operator's knowledge is used to program the robot, to monitor the welding process and to adjust the wire feed speed, weld current setting etc. accordingly.

9.1 Survey of robot applications

Materials handling

Ever since the early days of robotics, materials handling has been an important area for automation. Robots of all six major configurations are used for handling, including the moving of components or assemblies to and from transfer lines, the loading and unloading of machine tools and the stacking and packaging of parts. Robotic handling also includes the transfer of workpieces to and from dangerous and hazardous processes such as die-casting machines, furnaces and heavy presses.

The use of robots for materials handling can result in increased productivity due to:

- Shorter time cycles
- Reduction in work in progress
- Elimination of the need for personnel to work in hostile or unpleasant environments

Palletizing and stacking

There are many examples of robots being used for palletizing/depalletizing and the stacking or storing of parts. A Reis R625 robot being used in the palletizing mode is shown in Figure 9.1, picking up a cardboard carton from a conveyor and placing it onto a pallet. The configuration of this particular robot generates a large vertical working envelope; this enables cartons to be stacked at maximum height, while the robot still retains a large reach.

A robot can be programmed for palletizing by teaching it each individual move. Obviously this can be quite time-consuming and tedious for the programmer and most robot manufacturers provide dedicated palletizing software which requires only the width, length and height of the stack and the distance between the piles to be entered into the program. An

Figure 9.1 A Reis R625 robot placing a carton onto a conveyor (*Courtesy Reis Robot Ltd*)

Figure 9.2 A KUKA robot stacking door panels (*Courtesy KUKA Welding Systems and Robots Ltd*)

illustration of a KUKA robot carrying out the repetitive task of stacking/storing car body panels is shown in Figure 9.2.

CNC machine tool loading

Manually loading and unloading CNC machine tools is a monotonous and arduous task often performed in dirty and noisy working conditions. Use of a robot or robots gives the advantages of:

- Lower turnover of personnel
- Shorter cycle times
- Increased utilization of machine tools, e.g. three-shift operation

It is normal practice to group the machines around the robot with conveyor systems transporting the parts to and from the robot-served production cell. Figure 9.3 shows a Unimate Puma 700 robot loading and unloading parts between two CNC lathes on an exhibition stand. This shows two turning machines, but the choice of equipment for such a production cell is dependent upon the part or family of parts that is being manufactured and could include grinding machines, drilling machines and deburring machines as well as lathes and milling machines.

Another example of machine tool loading is given in Figure 9.4 where a Reis R625–40 kg electric robot is loading and unloading turbine components for the aerospace industry into a creep-fed grinding machine.

Figure 9.3 Unimate Puma 700 robot loading and unloading machine tools (*Courtesy Staubli Unimation Inc.*)

Figure 9.4 A Reis R625–40 kg robot in the palletizing mode (*Courtesy Reis Robot Ltd*)

Investment casting

In investment casting exact replicas of the finished component are first made in wax from accurately machined metal dies. They are then positioned together to form a 'tree' of parts, carefully dipped several times into a slurry of ceramic material and agitated to ensure an even coating. The tree is then placed in a heating chamber where the wax is melted out; this results in a hard but brittle ceramic shell containing a perfectly formed cavity into which the molten metal is poured. This process is sometimes referred to as the 'lost wax' process.

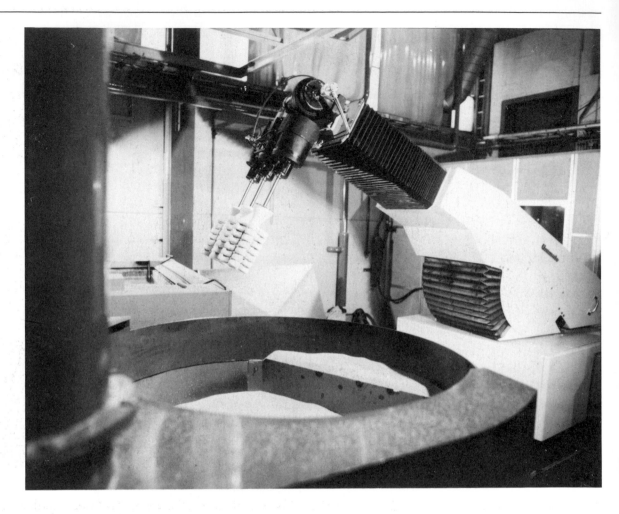

Figure 9.5 A Unimation robot dipping wax 'trees' into ceramic slurry as part of the investment casting process (*Courtesy Staubli Unimation Inc.*)

Figure 9.5 shows a Unimation robot dipping a fragile 'tree' of parts into the ceramic slurry prior to loading into the furnace for heating. This is an application for which robots are well suited since it is a repetitive task requiring careful execution in unpleasant working conditions.

Assembly operations

Perhaps the greatest unexploited area for the use of robots is in parts assembly operations. The recent development of vision systems, tactile sensors and part-feeding devices is now enabling many labour-intensive assembly tasks to become robotized. Vision systems can identify parts and also guide the robot to pick up randomly-placed objects from conveyor belts or feeders. Once the part has been positioned in a fixture or in the assembly, the robot can change its end-effector to accommodate the required tool for the task to be implemented. This is achieved by having a dedicated tool-changing station, or indexing a multi-tool end-effector on the robot wrist.

Figure 9.6 A team of Unimate Puma robots assembling cylinder heads (*Courtesy Staubli Unimation Inc.*)

The automobile industry is increasing the use of robots for mainline and sub-assembly operations. This is very evident in cylinder head assembly: teams of robots are used to assemble the complete cylinder head, including such components as valves, cotter pins, seals and valve springs. One such line is shown in Figure 9.6 where cylinder heads are being assembled by five Unimate Puma 560 robots. The system handles batches of three different types of cylinder heads in the order:

Insert valve
Insert spring seal
Fit stem oil seal
Load spring, washer and collet
Compress the springs and seat collets

The first task is to insert the inlet and exhaust valves. This is done by two robots each equipped with sensitive grippers which detect correct assembly and reject bent or oversized parts. At the second workstation two robots assemble the spring seat and oil seal over the valve stem guide. The assembly force reaction is taken by docking onto the valve stem for

Figure 9.7 A Reis V15 robot assembling automobile transmission cases (*Courtesy Reis Robot Ltd*)

anchorage. Finally, a single robot loads an assembly consisting of the spring, the top washer and the retaining collets over each of the eight valve stems. This system has replaced a boring and highly repetitive human operation with automation that assembles 103 cylinder heads per hour, with consistent quality and reliability.

Robots are also being used for the sub-assembly of automobile parts. Figure 9.7 shows a Reis V 15 robot assembling a transmission case complete with gears, shafts and bearings. In this particular case the robot actually looks for features on the aluminium casing, such as ribs and references component positions from these. This has the advantage that the cost of fixturing used is greatly reduced.

Spot welding

Spot welding in the automobile industry was one of the first applications of industrial robots. This was due to several reasons:

- The task was labour-intensive, boring and repetitive.
- The early hydraulic robots had the capacity to carry the weight of the spot-welding gun, up to 100 kg, and, with this load attached, perform with speed and accuracy.
- The reach of the robot and the quality of the welds were far superior to human performance.

Since those early days, spot-welding guns have become more compact and lighter. Due to this reduction in size and weight, electric robots are now used. These have the advantages of greater accuracy and reliability, are faster in operation and, most important, they require less operating space on the shop floor.

The process of spot welding is quite simple. Two copper electrodes clamp the steel sheets or body panels together and a low voltage is pulsed through the joint. A high temperature is generated at the interface of the two parts; this is caused by the high electrical resistance offered by the metal to the current passing through it. The current can be as high as 1500 amperes. When the current is turned off, the electrodes remain clamped long enough for the weld to cool; the rate of cooling is enhanced by using water-cooled electrodes.

A typical spot-welding cycle takes the form of:

Move electrodes to position
Clamp the parts together
Weld
Hold while the weld cools
Release
Move to the next spot-welding position

Figure 9.8 shows a spot-welding gun attached to a Unimation robot welding sub-assembly. The transformer, clamping mechanism and electrode cooling equipment can be seen clearly.

There are two different approaches for orientating the robot to carry out the task of spot welding: (a) as in Figure 9.8, the robot, with gun attached, moves to the component which is located in a fixture; (b) the part to be spot-welded is held by the robot which moves it to a welding machine. Obviously the main consideration here is the size and weight of the workpiece – it would be impractical to heave car body panels around in such a manner. However, for small panels and sub-assemblies it may be easier to take the workpiece to the welding station rather than attaching a heavy welding gun to the robot.

It is common in the automobile industry to increase the production rate of spot welding car bodies by using two robots situated either side of the track, as shown in Figure 9.9. However, the number of robots employed can be increased by mounting robots overhead, as shown in Figure 9.10, where three KUKA robots are spot welding the rear ends of car bodies.

Figure 9.8 A spot welding gun attached to a Unimation robot (*Courtesy Staubli Unimation Inc.*)

Figure 9.9 A series of Cincinnati T^3 700 electric robots spot welding on an automobile production line (*Courtesy Cincinnati Milacron*)

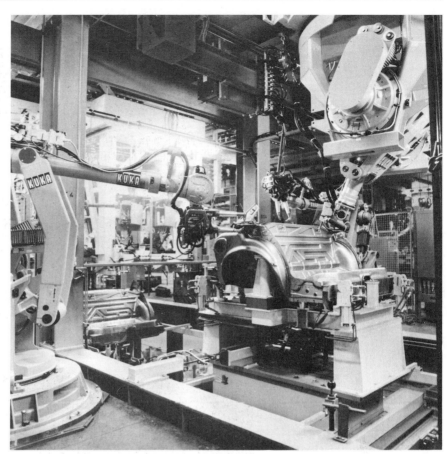

Figure 9.10 Three KUKA robots (one overhead-mounted) spot welding the rear ends of car bodies (*Courtesy KUKA Welding Systems and Robots Ltd*)

Inspection and test

The use of robots and associated vision systems for the inspection and testing of sub-assemblies and finished manufactured items has increased since the mid-1980s. The reason for this growth is that reliable, high-resolution vision systems have been developed to work in conjunction with advanced robot software. Also there has been considerable advancement in laser technology and associated software, enabling accurate measurement to be taken in three dimensions. This has opened the way for the automatic detection of defects and the checking of the shape and size of manufactured components against computer models. An example of this is given in Chapter 4 where the automatic inspection of automobile bodies is discussed.

In general most visual inspection systems are based on replacing the component to be inspected with a detailed software model. By using multiple cameras, a detailed view of the object is presented to the image process, the picture is analysed and cross-referenced to the computer software model. The benefits of using such systems are:

- Accurate and consistent results are obtained
- No high-precision inspection fixtures are required
- The system is independent of human error and judgement
- The inspection system can be fully integrated within an automated production system

Such a system is shown in Figure 9.11 where a high-resolution vision system is mounted on the gripper of a Unimate Puma 260. The robot is working in conjunction with a Unimate Puma 560 robot in the combined assembly, inspection and calibration of domestic television receivers.

Figure 9.11 A Puma 260 robot carrying out assembly, inspection and calibration of television receivers (*Courtesy Staubli Unimation Inc.*)

9.2 Flexible manufacturing systems

There is much in common between an automated production line and a flexible manufacturing system (FMS): both can be as simple or as complex as the production process requires. The governing principle of an FMS is that raw material fed in, by some suitable method, at one end of the system finally leaves the system as a finished article. Obviously not all industrial manufacturing operations can be adapted to such a system.

SCAMP (Six-hundred Computer Aided Manufacturing Process) was built, with the support of the Department of Trade and Industry, as a demonstration unit to show industrialists what can be done when CNC machines are linked together. When the SCAMP line was built the technology involved was quite new and this was the first demonstration of FMS. Figure 9.12 gives an overall view of the system.

All parts are loaded manually onto pallets, using either a template for accurate component placing or a box mounted on the pallet (Figure 9.13).

Figure 9.12 The SCAMP production line (*Courtesy 600 Group Colchester Lathe Company*)

Figure 9.13 Box for mounting on pallet (*Courtesy 600 Group Colchester Lathe Company*)

To indicate the parts to be worked on, at each loading station the monitor shows a picture of the template beside which is the batch number and batch quantity for that particular run (Figure 9.14). As certain parts are required, the pallet holding those parts moves forward to a unit that pushes the pallet onto the inner conveyor system; this carries the pallet anticlockwise round to the cell that is calling for more of that particular batch run. The pallet is then pulled onto the outer conveyor and into a waiting area where it will finally reach the unload station, which is also robot operated. The robot can be separately programmed, in a small number of steps, to carry out a depalletizing routine.

On receipt of certain basic instructions or commands the system software automatically calculates the numbers of component parts and their positions on the pallets, even if they are stacked. This facilitates reprogramming and the control of operational cycle times.

When the components have been machined they are returned to a waiting empty pallet which then follows a palletizing routine that is programmed exactly as the earlier routine. When it is fully reloaded, the pallet is released forward to a pusher that pushes it into the centre conveyor system and returns it to its original position to be unloaded manually.

The SCAMP line, no longer a demonstration unit, is housed in a purpose-built building in Colchester, Essex, where it is producing parts for the 600 Group Colchester Lathe Company, working a three-shift system over 24 hours.

Figure 9.14 Load station (*Courtesy 600 Group Colchester Lathe Company*)

9.3 Arc welding

Plasma arc welding

The term 'plasma' refers to a gas, usually argon, that has been sufficiently ionized to conduct an electrical current. The basis for this change is the addition of heat to the gas, since if sufficient energy is added the temperature becomes high enough for the molecules of gas to deactivate so that the gas exists in its atomic state. If the temperature is further increased, the atoms will lose electrons to become ions. The gas then consists of a combination of positively charged ions and free electrons. This is known as the plasma state and it is the availability of free electrons that allows the easy passage of electrical current.

Plasma occurs to some degree in all welding arcs, but plasma welding exploits it to a far greater extent by constructing the arc to form a high-intensity collimated or linear arc stream. This can be further explained by comparing the shape of the plasma welding arc with that of a tungsten inert gas (TIG) welding arc. A typical TIG arc has a 60° cone angle (depending on arc gap) which results in the heat input to the weld varying inversely as the square of the distance between the electrodes and the workpiece. As a result, small changes in arc length cause large variations in implement area and consequently in the heat input per unit area. If the weld penetration and bead width are to remain constant it is therefore necessary to operate with short arc lengths, typically between 0.5 and 1.0 mm and 3.0 mm when using wire. In the case of a plasma arc, the cone angle is 6° so giving an almost uniform intensity of heat input over a wide range of arc lengths. Arc lengths of 4 to 10 mm are possible with minimal variation in penetration and bead width.

This greater tolerance of arc length variation is beneficial when considering machine or robotized welding of un-fixtured joints, where heat dissipation may cause a variation in work-to-torch distance due to distortion of the parent metal. In autogenous welding, alignment of arc with seam is more critical with plasma welding as a TIG arc will deflect a small amount.

Variations of plasma welding now exist covering welding currents from a fraction of an amp up to 400 amps, thus enabling a wide range of material thickness to be welded, i.e. from thin metal foil to material of 25 mm thickness.

As with pulsed TIG welding the power source is microcomputer controlled. Parameters such as peak current, background current, pulse time and modulation level are set and displayed on a VDU in menu form. Synchronization of peripheral devices such as the wire feed unit and the welding robot arm with the welding cycle is controlled by the welding program. With the ability to link several programs within the welding set, there exists the facility to program automatic changes in operating conditions during the welding cycle. The scope for joining dissimilar thickness of materials, different joints and positional welding is considerably enhanced.

Typical examples of plasma welding applications

- Fusion of aluminium transformer windings
- Butt welding of aluminium coils to make continuous strip
- Welding of high-temperature alloy steam turbine blades to the outer ring
- Sealing of welds at the crimped end of copper tubing for thermostats
- Welding of brass for musical instruments
- Welding of aluminium and steel tubing for office furniture

Precision pulsed TIG welding

Low-current pulsed-arc TIG (tungsten inert gas) welding is becoming widely used for welding applications which require low, localized heat input. This is particularly desirable in the realms of high technology where it is necessary to join thin and expensive materials at comparatively low heat inputs with minimum rejects. Pulsed TIG welding allows metals to be joined using a heat input as low as or lower than that of methods such as laser or electron beam, which are more costly and more difficult to robotize.

The welding can be selected to be carried out either in pulsed or unpulsed mode, and the controls enable the full weld sequence to be programmed.

The system uses digital electronics for precise definition of pulsing, timing and current functions together with a linear amplifier to give continuous control and monitoring of the output waveform.

Pulsing the arc gives a not unattractive 'fish-scale' appearance to the finished weld as well as optimizing penetration and minimizing heat input. Since the pulsed TIG power source gives square-wave pulses of extremely high consistency, it is necessary to present the work to the robot manipulator accurately, not only to equalize the heat input, but to ensure that the pulses overlap equally. Development work is being carried out to perfect an automatic seam-following system that will attach to the arc length controller (ALC), thus enabling the robot arm to compensate accurately for any 'run-out' in the weld seam.

Typical weld sequence using a robot

1 Robot moves to above the component which is located in a purpose-built fixture, designed if necessary to hold the component in such a way as to reduce any possible distortion by acting as a heat sink.
2 The robot sends an output signal to the ALC, which is located at the end of the robot arm, and the robot waits in this position until it receives an arc established signal after the execution of gas prepurge, weld strike and current upslope.
3 A stepper motor drives the electrode in the ALC until it touches the component surface.
4 There is a gas prepurge and a gradual increase in weld current (upslope) to avoid craters, burn-through or spots at the beginning of the weld seam.
5 The weld established signal is received by the robot, the full-current weld period commences, the wire feed (if used) is turned on and the robot moves the torch along the programmed weld seam.

6 At the end of the weld seam the robot sends an output signal to actuate the downslope and gas postpurge. Upon completion the robot moves away and sends a signal to the welding set to retract the welding torch to its home position in the ALC.

This method of welding is becoming increasingly popular as a fairly general means of joining stainless steels, mild steels and different non-ferrous materials. TIG welding has certain advantages for producing very accurate welds when automated or robotized. Some of these advantages are listed below.

- It uses a low frequency to initiate the arc so does not interfere with the robot controller's microprocessor and DC thyristor drives.
- The equipment is relatively inexpensive and does not require water cooling or expensive fixturing to compensate for component inaccuracies and slight runouts.
- TIG is a much faster process than plasma, which sometimes requires a second run to achieve the required joint.
- The equipment is compact and portable; this is an important factor when considering robotization, where the space available may be restricted.
- Modern precision TIG welding equipment, such as that designed and manufactured by Precision Systems, uses an arc length controller (ALC). This electronically controls the predetermined distance between the end of the electrode and the weld surface. When programming a robot to perform perhaps four passes over a weld joint using filler wire to build up a surface, the program need only be taught once since the arc gap voltage is measured and a stepper motor on the ALC, attached to the robot arm, moves upwards to compensate for the previous thickness of the filler wire deposition.
- 95 to 97 per cent of all micro-welding can be just as satisfactorily carried out using TIG without the extra complications of plasma.
- A wide range of materials can successfully be welded, such as various steels, copper, phosphor bronze, tungsten, molybdenum and some grades of aluminium.

Typical applications of pulsed TIG welding

- Welding bellows and diaphragms
- Sealing and encapsulating welds
- Seam welds in thin wall tubing and thin flat sheets
- Miniature battery 'can' welds
- Jet engine blade recovery
- Seal fin repairs and nuclear fuel rod ends

The adoption of robots for arc welding

The first-generation robotic arc welding systems were not an unqualified success, with much of the blame being unfairly laid on the robot.

In reality the limitations can mainly be attributed to poor system concepts and engineering, coupled with inappropriate management of such installations. The potential flexibility of the robot was often restricted to that of a dedicated device, due to the availability of only unsophisticated welding processes. Many of these first robotic welding systems under-performed as a result of inept management, organization and practices. Maximizing the performance of robotic welding cells requires as much detailed attention to the logistics of work flow as to pure welding technology.

The reasons for these failings may be summarized as follows:

1 It was perhaps unfortunate that seam tracking for robot applications was not commercially available from the inception of the robot manipulator. The absence of this facility imposed an unduly severe constraint on edge preparation to achieve the necessary joint accuracy. Thus the product immediately incurred a cost penalty as a consequence.
2 An immediate limitation was imposed on the configuration of the component to be robot welded, due to shortcomings in process facilities. For positional welding to be effective, automatic adjustment of process operating parameters is necessary in real-time in order to correct for changes in torch position, arc length compensation and seam tracking. Added to this, real-time control is not achievable without micropro-cessor control of the power source, which has only recently become available.
3 The majority of robotics applications to date have concentrated on comparatively low-added-value products with relatively conventional materials. Scope for a rapid return on investment, even with high volume throughput, is severely restricted with this type of application. For a wide range of manufactured products, material accounts for approximately 50% of the product cost. Therefore it might be argued that one hour spent welding titanium or stainless steel products is likely to yield significantly better return on the welding system than the same period operating with mild steel.
4 Operation and management of an expensive robotics facility requires a different approach from that of a general fabrication shop. Lack of attention to the high cost and high throughput potential can result in the investment being under-utilized, so increasing the unit cost.

It must be stated that these reasons should not be allowed to inhibit the future acceptance of robotics for arc welding. Recent technical advances can overcome some of the early limitations. Sound economic reasons still exist for the adoption of robotics systems in selected areas of application.

Both plasma and precision TIG welding, with their high-temperature constricted arcs, infrequent use of high frequency and greater tolerance of arc length variations, provide process characteristics more compatible with the requirements of robotic applications. In the welding of aerospace-type materials, including titanium and nickel-based alloys, with robotic systems that offer inherent flexibility both these types of welding are uniquely placed. With the advent of microprocessor-controlled welding sources, the last of the essential building blocks for fully automatic intelligent robotic systems is now available.

What is now required is a fresh and more circumspect approach in the selection of products for robot welding. Many potential and viable applications exist, particularly in the aerospace, nuclear and process industries.

An automated welding system

Westwood Engineering Ltd use a unique but simple automated welding system to produce a range of tractor-style rides on lawn mowers. The Westwood flexible welding system has now replaced the S100 shown below with a 600 FANUC Arcmate industrial robot (GMF Robotics Ltd), which is one of the latest generation direct-drive robots of Japanese derivation that have been designed specially for welding. The robot has six axes (degrees of freedom), a 5 kg load capacity and uses digital encoders for fast feedback of positional data, which allows better repeatability and smoother interpolation between points.

Figure 9.15 600 Fanuc robot model S100 linked to Westwood Automation transport system (*Courtesy Autotech Robotics Ltd*)

The method of make-up of the flexible welding system allows the line to be extended or shortened, depending on the batch size and space available, or to allow for any particular requirement of the customer.

Parts are loaded manually onto pallets which are made in 12 m and 15 m lengths; the pallet surfaces are drilled to mount the jigging and fixturing to

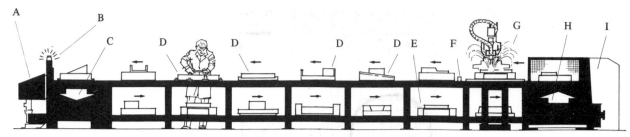

Figure 9.16 Westwood
Automation transport system
(*Courtesy Autotech Robotics
Ltd*)
A Control console offers three
mode options: automatic, semi-
automatic and manual.
Counters record all functions
and an illuminated schematic
display shows the positions of
the pallets
B Warning lights flash if the
welding torch has failed to arc
or if there is only one pallet
waiting to be welded
C Elevator lowers work from
the loading/unloading level to
the conveyor feeding the robot
D Loading/unloading station
E Work waiting to be welded
(up to 16 jobs can be carried)
F Cleaning station, where the
robot automatically cleans the
torch between welding
operations
G Welding station: short bolts
raise and position the pallet in
front of the robot. Proximity
switches identify the job to be
welded
H Lift raises the pallet prior to
presentation to the robot
I Robot controller

hold the components to be welded. The line is made up of load stations, lift
stations and welding station, of which the standard height is 1 m; it can be
built 10 cm, 20 cm, 30 cm or 40 cm higher to customer requirements. The
line width is standard at one metre; obviously the limitations of the line are
governed by its dimensions. The pallets can carry weights of up to about
200 kg which includes the weight of the jigging and fixturing.

When a part has been loaded onto the pallet the operator pushes a release
button that activates the section drives to transport the pallet around the
system; it eventually arrives at the welding system (Figure 9.16). Here four
pneumatically operated short bolts lock the pallet into position (Figure
9.17). The code on the pallet (Figure 9.18) is read by sensors and the
information is sent to the robot, which calls up, through the Robot Service
Request system, the appropriate pre-taught program to enable it to weld
the components on the pallet.

The line is controlled by means of a control panel (Figure 9.19) which
houses a programmable logic controller that controls all the inputs and
outputs for the line and integration between robot and the line.

The line is not limited to welding: it can be used as an assembly line
without the lower transport area, or as a mobile spray-paint transport
device for paint-spraying robots.

Figure 9.17 Pallet locked into
position for welding (*Courtesy
Autotech Robotics Ltd*)

Figure 9.18 Pallet recognition slide (*Courtesy Autotech Robotics Ltd*)

Figure 9.19 Central control panel for the Westwood Automation transport system (*Courtesy Autotech Robotics Ltd*)

9.4 Automatic bonding and sealing

Over recent years adhesive bonding has established itself as a reliable joining method for producing low and high strength joints. The major advantages offered by this technique are:

- Uniform transmission of forces onto the component
- The possibility of joining dissimilar materials
- The production of joints and seams which are impervious to gases and liquids
- Good damping properties of the bonded components
- The production of sheet-metal constructions which have smooth outer surfaces

These benefits are utilized in numerous progressive branches of industry, particularly in the automotive sector, where inflexible, dedicated bonding systems are being replaced by robotics facilities. The development of automatic application of adhesives and sealants has resulted in further advantages:

Figure 9.20 A KUKA robot manipulating a dispenser with two outlet valves which it twists into the correct position to apply two different adhesives to truck door panels (*Courtesy KUKA Welding Systems and Robots Ltd*)

- Higher production rates due to shorter processing times.
- Greater flexibility with regard to different parts and lot sizes.
- High positional accuracy of the adhesive-bonded joint.
- Good bonding properties of the component even around corners and along contours: the uniform application of the adhesive is achieved by using high precision dispensing equipment.
- The elimination of spot welds on hemmed seams. This has the advantage not only of making the seam more rigid but also that the risk of corrosion due to water seeping into the hem area is greatly reduced.

An example of this technology in use in the automobile industry is the automatic application of adhesives onto truck doors, as shown in Figure 9.20. A KUKA IR/161/15 robot applies two different types of adhesive to the outer door panel, i.e. an epoxy-resin-based material around the hem and a metal bonding adhesive in the area of the door handle. In order to implement this, the robot is equipped with a dual dispensing unit with two adhesive outlet valves. These are arranged at an angle to each other and twist into the correct position for the required application. The dispenser is mounted on the robot wrist and maintains consistent application of the adhesives regardless of any variations in their viscosity. Around the hem, the adhesive is applied in proportion to the robot's speed of travel; this results in an absolutely uniform adhesive bead even at the corners and along contours. To guarantee consistent quality of application, the volume and pressure of the adhesive are continually monitored throughout the operation.

9.5 Automated instrument component manufacture

Crompton Parkinson, a leader in their particular field of instrument manufacturing, are using an assembly robot to load and unload components into a group of machine tools, for the execution of various mechanical operations. The cell comprises a Bentley 32-ton power press, drilling, tapping and boring stations, automatic tool changing, component magazines and an indexing table. The layout is shown in Figure 9.21.

Two different parts are handled in the cell, clear plastic panels for the fronts of meters and small steel pressings.

The plastic panels are fed 50 at a time into each of five magazines positioned on an indexing table. A 600 FANUC A model 0 robot, fitted with flexible pneumatic grippers, picks up an individual panel (Figure 9.22) and loads it into an automatic vice, where it is clamped. The centre of the panel is then drilled and air-blasted to remove any swarf. Upon completion of these operations the panel may be bored, tapped and air cleaned according to the required specification. The robot then deposits the panel into an empty space in the magazine and proceeds to repeat the process on a new panel.

Alternatively, the robot can be used for loading and unloading the power press. The robot end-effector is changed at the automatic tool change station and then picks up a steel pressing from a magazine on the indexing table. The robot positions the component in the power press (Figure 9.23) and withdraws; the press is automatically actuated and, on completion of

Figure 9.21 The initial cell working plan (*Courtesy Crompton Parkinson Instruments Ltd*)

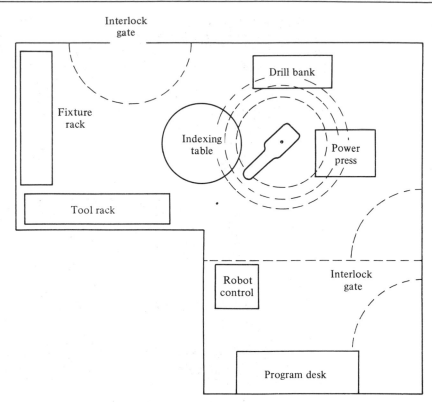

Figure 9.22 A magazine holding components is unloaded and loaded for drilling and tapping by the robot (*Courtesy Crompton Parkinson Instruments Ltd*)

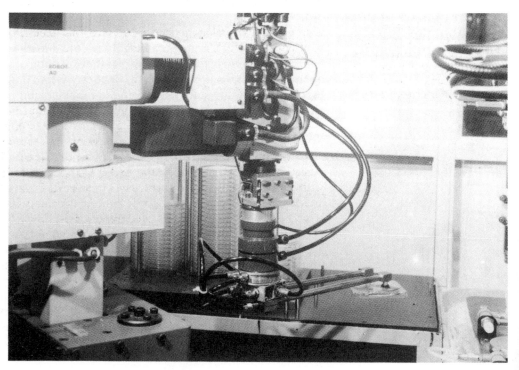

Figure 9.23 The robot loads components into the 32-ton press – this highlights the repeatability of the robot and the flexibility of the cell when linked with different peripheral devices (*Courtesy Crompton Parkinson Instruments Ltd*)

the press-stroke, the robot removes the pressing and drops it into a storage bin via a chute.

This flexible cell was designed, manufactured and installed by the company's own team of engineers.

Factors considered in the feasibility study

1 The parts could be transported to the workstations by 'hard automation': this would involve the use of conveyors and dedicated handling units. On reaching the pick-up point, the component would be orientated into position by suitable guide devices. This in turn would activate signals to enable the dedicated pick-and-place units, at each station, to pick up the component and load it into the machine. The handling of components would be by vacuum cups or mechanical grippers, depending upon the component shape.

2 Alternatively, a pick-and-place unit could be installed that would be of sufficient size to serve all four machines from the conveyors and presentation magazines. Applying this system would involve extensive electrical and mechanical modification to the existing machines.

3 The existing machinery could be easily and effectively modified and renovated to interface with a GMFANUC assembly robot; this would also allow the use of existing tooling without many major modifications.

4 The total cost of capital outlay was considered and a sensible payback period established.

5 The adopted system was required to be sufficiently flexible to accept a variety of components.
6 The cell might be required to expand to accommodate another robot at a later date.
7 Operators, programmers and maintenance engineers would need to be trained.
8 The development time required to get the cell fully operational would need to be allowed for.

After twelve months of production the robot cell was expanded to increase the efficiency of the equipment and also to minimize operator attendance. It can be seen from Figure 9.22 that this was achieved by using the existing component tooling plates mounted on a Camco Indexing table. The table was interfaced to the robot, so that the programmed output signals indexed the table when required, thus enabling the cell to be operational outside normal working hours.

Problems

1 What are the advantages of adhesive bonding over other joining methods? Why is this an ideal application for robots?
2 Sketch a layout for a robot cell for loading/unloading three machine tools with workpieces from an input conveyor. Describe the robot that you would use and give reasons for the choice of this type of system.
3 Explain the term 'palletization'.
4 Robots are increasingly used for assembly operations. What problems may be encountered in this application and how might they be solved?
5 Explain the process of spot welding and why this has been a successful application area for robots.
6 What are the benefits of the use of vision systems for automatic inspection?
7 Explain, with the aid of diagrams, how pallet recognition is achieved on the Westwood Automation line.
8 Explain the acronym SCAMP and briefly describe the basic functions of this FMS line.
9 How can a gas be converted into a plasma state? Give typical examples of plasma welding applications.
10 List a typical robot weld sequence for pulsed TIG welding. Why is this method of welding becoming increasingly popular?
11 What factors contributed to the failure of welding installations using early robots?
12 Outline the alternative methods considered by Crompton Parkinson for automating their instrument component manufacture.

10 The future of robotics

10.1 Current developments

Robots and their associated equipment have progressed beyond the first deaf, blind and dumb generation: the third generation of robotized systems employ vision, tactile sensing and communication with other systems to form the initial realization of expert systems, i.e. learning from experience.

Robots today are being developed and used in areas which would previously have been unimaginable. They include four-legged robots for the transportation of lumber in the USSR and Canada and for the deployment of equipment across marshy and boggy terrains. A sheep-shearing robot developed in Australia has a tactile clipper attached to the arm of a robot and hovers just above the skin of the sheep; the robot is programmed from a computer-based model of a live sheep. A robot for playing snooker uses vision, tactile sensing and an expert system to determine game strategy and how to play the cue ball.

Advances in vision and laser systems have made possible the random selection of components, the automatic monitoring and sensing of welded joints, detection of seam position and weld quality control. Cincinnati Milacron have now developed a robot for laser cutting, welding, heat treating etc. in which the beam is actually directed through the robot arm (Figure 10.1). This has the advantage that the laser beam can operate in any direction, making it much more flexible than conventional laser machines.

There have also been many advances in robotics in the medical field for body exploration and surgical operations.

The capacity of robots and robotized systems is being developed in many countries throughout the world. Some indication of the scope of current research projects is given by a brief outline of work being carried out in the USA and Japan.

Projects in the USA

Three-dimensional sensing

The development of tactile human-type hands which can operate very fast to grasp and manipulate an object of any shape.

Model-based vision systems

The outline of an object is compared with a three-dimensional model held in the computer database. It is intended that a robot will use a series of

Figure 10.1 Laser beam
directed through the robot arm
(*Courtesy Cincinnati Milacron*)

inspired guesses and comparisons to orientate itself correctly to retrieve a
randomly positioned object from among other objects in a confined space.

The development of mechatronics

This is the integration of computer control directly with the actuators and
sensors of the robot. This will give the robot an element of direct feedback
control over response to force applied to the manipulator: for example,
robot systems cannot at present respond quickly enough to stop the wrist,
hand and/or manipulated device from being destroyed in a collision.

Geometric reasoning

The ability to find collision-free paths between obstacles and to place or
pick up an object among a crowd of other objects in a cluttered
environment.

Communications

Intercommunication between machines, computers and ancillary devices,
i.e. a local area network for communication between processors and
interfaces. One such development is General Motors' Manufacturing
Automation Protocol (MAP) system.

Projects in Japan

Continuing development of conventional robot arms and applications:

- The design of manipulators with more than six degrees of freedom
- The manipulation of robots in confined environments
- Mechatronics
- Tactile sensing for deburring systems
- Windscreen insertion
- Computer disc drive assembly
- Watch assembly

Development of robot systems intelligence:

- Automatic inspection/disassembly of valves in nuclear power plants
- Automatic inspection in undersea exploration
- Recognition and inspection of objects in poor visibility

10.2 Advanced robotics*

The following comprehensive description of advanced robotics has been given by Ron Egginton, Advanced Robotics Programme Coordinator at the UK Department of Trade and Industry:

Advanced robotics represents a revolution in the development of robotics. The familiar industrial first or second generation robot is confined to a fixed point, with a limited scope for largely repetitive actions, and a limited awareness of the environment it operates within, or understanding of the task it is 'mindlessly' undertaking. By contrast, the advanced or third-generation robot is a much more exciting animal. Rather than being fixed in position, the advanced robot is mobile, whether by wheels, tracks, legs or other means. Instead of being purely repetitive in operation it is capable of applying artificial intelligence and decision-making capacity to a range of tasks which for the operational instructor would involve a high level of control, or even voice command. To work effectively, the advanced robot needs to be aware of its environment and for this purpose it needs to be equipped with a range of tactile, ultrasonic, infra-red or audio sensors. Interpretation of the sensor information is a key computing requirement. The advanced robot could be termed an autonomous robot able to operate independently.

Advanced robotics can therefore be looked upon in its fully developed form as providing a combination of human and mechanical capabilities without many of the limitations or constraints of either. Advanced robots should be mobile enough for deployment to their worksites, in whatever environment they may be, without the life support or remote power supply of humans or existing mechanical devices.

To achieve this they must have the primary senses of a human and the means of processing, storing and acting on the data they provide, together with the strength, mobility, repeatable accuracy and endurance of machines. It will be several years before such robotic devices are fully

*The discussion in the following pages closely follows (1) and (2).

developed but it is only by considering what they will offer ultimately that the opportunities for short-term development can start to be seen.

Projects with robotic-type devices are being developed worldwide and aim to create the third and fourth generations of robots, whose attributes include advanced sensory capabilities that permit the use of sensor-stimulated arm/hand movements to solve certain levels of unstructured disorder, and also the ability to communicate and integrate with the environment and to learn, store and manipulate a fund of human knowledge (expert systems).

The areas of application in which these developments are taking place include:

Tunnelling
Civil engineering and construction
Underwater operations
Nuclear technology
Fire fighting and emergency rescue
Space operations
Medical health care
Agriculture
Domestic and leisure applications

Advanced robotics in space

To operate in space, special-purpose automated mechanisms are required. Early examples of robot-like systems were *Lunar Surveyor* in 1967 and the *Mars Viking Lander* in 1975.

More recently the need to minimize risk to personnel and the cost of manned operation have led to the development and use of the Remote Manipulator System (RMS) on the US Space Shuttle. In addition, NASA foresees that in the post-*Challenger* era there is likely to be a reduction in the number of space initiatives. This will emphasize the need for automation and robotics as a means to achieve the necessary improvements in 'on orbit' productivity and efficiency.

In the future, space robots will perform tasks that are impossible or unsuitable for humans, e.g. remote servicing of satellites in high orbit. In other cases, a robot will simply constitute a better option, on grounds of cost and technical efficiency, than the equivalent special-purpose auto-mated system. Robots will also be a cheaper way of doing a job that would otherwise require an astronaut – it is very expensive to launch and support men in space.

Specific features of space robotics systems

A number of constraints and requirements that are particularly crucial to the use of robotics in space are unlikely to be encountered and tackled to the same extent in other areas of application. These include the following.

Zero gravity

This is unique to space and no true performance tests are possible on Earth. Special control and lubricant systems are required.

High vacuum

Materials limitations, the need to avoid contamination of other mechanisms, heat transfer limitations and temperature extremes are major considerations.

Damage from the environment

Space radiation and particle bombardment from meteorites or man-made space debris.

Special communications requirements

Significant time delays have an important influence on control loop concepts and on the need for intelligent and autonomous remote operations. Special-purpose, high-capacity, fast real-time tele-presence and telecommand systems are needed.

Self-diagnosis and repair

Maintenance requirements must be either very low or zero. The high level of reliability required includes, for instance, ability to detect faults automatically and restructure the control system to work in the presence of component failure.

Light weight

Reliability based on bulk and mass is not an option because of launch constraints: this imposes the need for careful selection of materials for equipment and control systems.

Moving frames of reference

Space robot vehicles may be in motion, and may also be attempting to capture objects that are spinning arbitrarily. Hence there are consequent design requirements for object recognition and for the control of motion.

Remotely deployed tool set

There may not be the option to select tools from the human operator station. In this case a remote and autonomous tool selection system may be required.

Long-term technical developments required for robots to operate in space

Actuators: *direct drive motors*

These are motors capable of delivering sufficient torque to drive the robot without the aid of a gearbox. This simplifies the problem of servo-control by eliminating the effect of backlash and actuator compliance. Piezo-electric motors are recent innovations which use the piezo-electric effect, rather than electro-magnetic fields, to derive the torque.

Space vehicle propulsion units

These must be included in the set of actuators: the robot that is mounted on a free-flying device will exert forces and torques on the base, which in turn will affect the altitude and orbit control system of the craft. The most advanced type of actuator envisaged would be 'intelligent' to the extent of providing for such functions as safety monitoring, self-diagnosis and inter-joint communication.

Manipulators

Flexible arms are a requirement of space robotics, where payload launch restrictions require large robotic devices to be lightweight. This makes the control of such devices much more complex. Long-term technology challenges include 'legged' motion, where the principal challenge is the analysis of gait to provide a stable motion. Advanced robotics will provide a much greater degree of dexterity in end-effectors, possibly emulating the human hand. Alternatively, two or more cooperating arms may be used as a dextrous end-effector.

Sensors

A robot in space may be required to be mobile – for instance, a servicing robot mounted on a free-flying craft and requiring sensors to measure the status of the base. This implies six degrees of freedom (three for position and three for orientation). A major problem to be overcome is the lack of a fixed reference point.

Long-term technical challenges include non-contact position sensors. These could be optical devices, needed for determining displacements due to flexing. Such sensors are advanced in that they require their own control systems to maintain pointing accuracy. Image-processing systems are used to translate the picture from a vision system into a form that is usable by the computer control system. This represents a challenge because of the large amount of data that needs to be processed in real-time. Often the arrangement of lighting in a space robotics application is very restricted. The basic picture from the camera might be improved by using a computer image enhanced by simulation of shadow effects from different lighting angles. This is a good example of the interaction of simulation and the 'real world'.

Pattern recognition is another key artificial intelligence aspect of an advanced robot. Its purpose is to take raw sensor data and organize it into shape information for use by the intelligent controller.

Controllers

The controller is the heart of a robotics system and involves many different key areas of research. Long-term technical challenges, mostly in the field of artificial intelligence (AI), aim to provide the degree of autonomy that is required to meet the demands of the man–machine interface. These include:

Geometric reasoning – the process of planning a path for the robot to get from one position to the next without colliding with anything in the environment. Although this problem is at a fairly low level in the overall hierarchy, it is still one that is not particularly well understood and is the subject of much current research.

Knowledge-based systems – a key technique in artificial intelligence. The essence of this technique is that decisions are made not at the level of the programmed algorithms, but rather on the basis of a sequence of statements related by rules (a knowledge base). The truth or falseness of one statement may, in combination with other statements, trigger off a new rule, which in turn will assert new statements. Thus, the rules encapsulate the 'knowledge' which defines the logical relationship between statements. This leads to a reasoning process which is able to cope with complex unstructured systems, such as route planning in an unknown environment.

Advanced robotics in underwater operations

The areas where advanced robotics may be used in underwater work include bottom survey, cleaning and inspection, and repair and maintenance of equipment. The specification for an advanced robotic device for bottom survey might include:

- Visual and sonar search for the inspection of pipelines, cables, miscellaneous objects, minerals and underwater food resources.
- Monitoring of salinity levels, temperature, current and tides, pollution and specific measurements, e.g. oxygen content.
- Study of water conditions and marine life in deep water and under ice caps.
- Charting surveys for undersea structure sites, pipeline and cable routes and waste disposal sites.
- Plotting lake bed contours.

In reality it would not be viable to incorporate all of these functions into a single vehicle. The vehicle configuration may be profoundly affected by a number of variables including:

- The application.
- The specific task to be undertaken.

- The environmental conditions in which the task is to be performed.
- The duration of the operations.
- The type and magnitude of the data to be processed: for example, the tasks for which the vehicle is specifically intended will demand varying degrees of stability, accuracy of position and altitude and varying speed capability.

A further and obvious example is the depth of operation. The depth capacity of the pressure vessels and of the buoyancy material will affect the vehicle's size and hence its drag characteristics and thus the energy required to complete a given operation. In addition, the operational depth could preclude what might otherwise be an acceptable power source; for example, the use of hydrocarbon fuels plus oxygen incurs severe penalties in terms of power density at significant depths.

These examples consider just two of many important parameters: it would clearly be difficult to design a vehicle that would be capable of performing all of the identified tasks while retaining viable proportions.

An advanced robot designed for sea-bed survey tasks would be a free-swimming mid-water device, carrying sensor equipment, that could be used for a range of duties. The tasks already mentioned might form a basis for a vehicle specification but the list is by no means exhaustive. A versatile vehicle could allow for numerous additional tasks to be considered. For example, if contact with a structure or the sea-bed is allowed, lightweight sea-bed samples or specimens of marine growth can be taken. In-situ analysis of some sea-bed materials is feasible.

Of course, operation is not limited to the proximity of the sea-bed; many types of work can be carried out at shallow and intermediate depths. Some examples are:

- Marine sciences research
- Fishing industries support/search/monitoring, herding shoals
- Location of icebergs which may threaten drilling platforms
- Under-ice surveys including measurement of ice thickness

An evaluation of the feasibility of an autonomous AR survey vehicle identifies five main considerations:

1 Autonomous vehicles already exist with programmed routines, obstacle avoidance and limited through-water data link capabilities. Since the technology used in these vehicles is not extraordinary by present standards, it is clear that major improvements can be made by the use of advanced techniques.
2 The major problems to be resolved are in the areas of advanced computer control, environmental interpretation, power/energy systems and long-distance navigation. The overall capability of the device depends to a great extent on the amount of on-board processing power available and the use of effective systems to provide intelligent control.
3 Subsystems for perception and interpretation of the surrounding environment are a prerequisite for autonomous operation, and this offers scope for considerable future development.

4 Development of power and energy storage systems is necessary to reduce the size and weight for medium and long mission durations.

5 Accurate long-distance navigation in an unstructured environment is barely feasible at present but the resolution of this problem would be of great benefit for numerous reasons.

Potential benefits of an advanced robotic survey device

Existing sea-bed survey techniques involve the use of towed fish or manned submersibles, both of which are costly to operate. The main expense arises from the need for continuous use of a surface ship; an advanced robotic device would bring considerable cost savings by requiring only intermittent use of a small vessel. In fact any such robotic device would frequently be deployed from a base on shore or from fixed installations at sea, making its own way to the worksite.

Dispensing with an umbilical cable provides further benefits. Cables constitute a severe hindrance to most sub-sea vehicle operations, and preclude some completely: under-ice surveys, for example. Cable-handling systems require deck space and add to costs and, for deep sea operations, the size and cost of cables and winch systems can become predominant factors.

Potential spin-offs

Many spin-offs are likely from development of the numerous subsystems which would make up the vehicle, not only for sub-sea use but also for terrestrial and aerospace applications. Intelligent control techniques, environmental interpretation, power systems and long-distance navigation systems are typical subsystems which offer tremendous scope for innovation and development, and a free-swimming autonomous machine would provide an ideal testbed for proving the increasingly complex technology.

10.3 Summary

The future development of robotics involves many exciting areas of development in applications which only a few years ago would have seemed extraordinary or impossible. However, to produce the performance that will meet the challenges that lie ahead the major requirements are the integration and intercommunication of sensing and vision systems with knowledge-based learning and the development of high-speed, lightweight and flexible manipulators.

References

(1) 'Feasibility Study for the UK Department of Trade and Industry on Advanced Robotics for Underwater Work'. VEG (CIRIA), July 1987.

(2) 'A Feasibility Study on Advanced Robotics for Space', prepared for the Department of Trade and Industry, Mechanical and Manufacturing Technology Division, Harwell Laboratory, January 1988.

Index